worlds without end

also by John S. Lewis

MINING THE SKY

RAIN OF IRON AND ICE

WORLDS WITHOUT END

the exploration of planets known and unknown

JOHN S. LEWIS

Helix Books

Perseus Books
Reading, Massachusetts

Many of the designations used by manufacturers and sellers to distinguish their products are claimed as trademarks. Where those designations appear in this book and Perseus Books was aware of a trademark claim, the designations have been printed in initial capital letters.

ISBN 0-7382-0170-7

Library of Congress Catalog Card Number: 99–64872

Perseus Books is a member of the Perseus Books Group

Cover design by Robert Dietz
Text design by Dede Cummings
Set in 11/14 Electra by Pagesetters, Inc.

1 2 3 4 5 6 7 8 9-   -02 01 00 99

Find Helix Books on the World Wide Web at
http://www.perseusbooks.com

To my parents,

John Simpson Lewis, Sr.

and

Elsie D. Vandenbergh Lewis

for giving me love and encouragement,

teaching correct principles,

and letting me choose my own path,

and to three recently departed colleagues and friends:

Carl Sagan,

for his contagious love of the Universe, and for always asking why;

Jurgen Rahe,

for his incisive intellect and warm humanity,

and

Gene Shoemaker,

for his voracious curiosity and unquenchable enthusiasm.

Without them, the worlds are poorer places

CONTENTS

worlds without end

INTRODUCTION

PERHAPS THE MOST EXCITING legacy of scientific research in this decade is the discovery of dozens of planets revolving about other stars. The fact of their existence has begun to sink in; their implications for life and intelligence have scarcely been explored. But at last we know the truth: as a few visionaries long ago anticipated, our Solar System is not alone.

Philosophers, both natural and unnatural, have speculated endlessly since ancient times about whether other worlds exist, how common they might be, whether they are "like Earth," how "Earthlike" they need to be to allow the origin of life, whether they might in fact harbor life, and whether advanced, intelligent life forms (i.e., similar to human philosophers) might be found on them. Some religions have formed powerful, deeply entrenched beliefs with respect to these questions, some of them even based on Scripture. Most answers and predictions, especially in the Judeo-Christian tradition, seem more firmly linked to speculation than to higher authority, although there are interesting exceptions. Beginning with the Greeks, another class of speculation, loosely termed science fiction, has also suggested a wide range of answers.

Fortunately, in recent years, both observation and theory have come to our aid, reducing the need for speculation, although not doing away with it altogether. In the twentieth century, astronomical and spacecraft explorations of the heavens have begun to answer many of these great questions. In our search for answers, we shall embark first on a Grand Tour of the Solar System, seeing with our own eyes the great variety of planetary bodies in our backyard. The chances of finding more planets in our Solar System are slim, but the ones we know about are provocative enough. As we compare these bodies with one another, certain systematic trends become apparent, revealing genetic and evolutionary forces that cause these bodies to fall into several loose families; but when all the nine planets and the seven planet-size satellites are compared, the troublesome thought arises that the sample of planetary bodies in the Solar System may be incomplete.

When we examine the 4.6-billion-year-old planets in our Solar System, we must struggle to distinguish the features that are due to conditions of origin from those due to divergent evolution after formation. Despite the remoteness of the time of formation, we have many insights into the events surrounding the origin of this family of worlds. We have, for example, powerful evidence from the study of the Solar System that planets are born along with their parent stars. We also have an exciting and rapidly evolving perspective on planetary and stellar origins from astronomical observations of places in our galaxy where stars are now being formed. We know from these observations that young stars exhibit a variety of violent and exhibitionist forms of behavior that shape their forming worlds, attracting the attention of terrestrial astronomers.

To understand a world and assess its suitability for the origin and proliferation of life, we need to know what it is made of, how it evolves, where it is located, and what kind of star it orbits. Each of these factors is of enormous importance. Many plausible combinations of these factors are not represented in our Solar System: the planets that orbit our Sun are not by any means the only kinds of planets that are possible — or even common — in the Universe.

Composition, we find, plays a major role in governing both the conditions on the surface of a planet and how the planet may evolve. The rocky planets, which are predominant in close proximity to a star, are made of several classes of material that have different conditions of formation. From high temperatures to low temperatures (in other words, in order of distance from the central star) these materials are refractories (hard-to-evaporate solids like firebrick, rich in calcium, aluminum, and titanium), metals (impure iron-nickel alloy similar to iron meteorites), silicates of iron and magnesium, compounds of sodium and potassium (feldspar) and sulfur (iron sulfides), oxidized iron, and chemically bound water (such as in clays). Rocky planets form a sequence, in which the growing young planets most distant from their star tend to have more of the latter materials. Liquid water, the essential medium for life, emerges during planetary evolution. Thus, though the fate of a planet depends both on its conditions of origin and its evolutionary path, the path it follows in its evolution is strongly affected by its composition, which in turn is governed by its mode of origin. Cleanly separating the two effects is not generally possible.

The evolutionary behavior of a planet, especially the way it responds to internal heat produced by radioactive decay, is also very sensitive to how big the planet is. Adding a little mass can cause striking qualitative changes. Likewise, adding a little heat from other sources, such as tides or early heating by a young star, can make evolutionary paths diverge. Bodies of different sizes cooked from the same recipe may end up very different in appearance. Each planet has a distinctive suite of surface rocks and minerals that reflect its bulk composition and history, and each releases volatiles that may react with the surface, or accumulate to form an atmosphere, or condense into oceans and frosts, or escape into space. The inorganic evolution of rocks, minerals, gases, and liquids paves the way for the origin of life.

The menu of available chemicals on a planet depends on its temperature. Very close to a star one finds planets whose surfaces bear high concentrations of refractories and metals. The more volatile metals, sulfur, and water are increasingly abundant at greater distances from the Sun. Further, gases escape readily from Mercury-size planets—they are too little to compete with their sun for control of gases. On all small planets, water and other gases may be released early, but they escape with ease, derailing the course of planetary evolution.

Not all small planets start out as small planets. Impacts of planet-size bodies can hurl off splashes of magma and chips of shattered rock that are virtually free of dense core material. The Moon is apparently just such a body. The Moon apparently formed as the splash from the impact of a Mars-size planet on the early Earth. Such collisions are probably not rare.

In addition to ejected chips, giant impacts can also leave planets that have been partly or wholly denuded of atmosphere, oceans, crust, and even mantle. One of these broken planets is Earth. What would Earth be like if such an impact hadn't happened?

Quite aside from random catastrophic events such as giant impacts, planets change with time, and do so in a manner that depends on the mass of the planet. A family of bodies with identical initial composition will follow different evolutionary paths depending on how big they are. If Earth were a little smaller, what would it turn into? And if Earth were a little larger . . . or even twice as big? Just how Earthlike (Edenic) is Earth? How

Earthlike are our larger and smaller siblings? How Earthlike must they be to harbor early life?

Other modest initial differences, such as exact distance from the Sun and precise original composition, may also have profound long-term consequences. Mars and Venus are examples of bodies that are compositionally distinct from Earth and at different distances from the Sun. Their evolutionary paths, both involving loss of water, have led to wildly divergent present-day conditions.

Beyond the rocky inner planets, our Solar System shows us examples of two kinds of giant planets, the Jupiter/Saturn and Uranus/Neptune types, which differ greatly in their content of icy and rocky materials. Planets of these types may look very different when exposed to different external temperatures. But even more dramatic differences follow from these planets' many different sizes, smaller and larger than those in our system. Even the eight extrasolar planets seen to date, all of which belong to these categories, illustrate a wide variety of sizes, distances, and temperatures. Are there regions in the atmospheres of these giants that might provide stable ecological niches for any conceivable form of life?

Superplanets, Jupiter-like bodies much larger than Jupiter but too small to become stars, can glow faintly with their own internal heat. These "brown dwarfs," though faint and elusive, have been tracked down by diligent astronomers. But what we see is certainly not the whole story. What conditions prevail deep within their atmospheres? Is there anywhere in a brown dwarf that life might hide? And what of conditions on the surfaces of satellites that orbit around brown dwarfs?

Seven world-size members of our own Solar System orbit not around the Sun but around planets. These local examples give some hint of how wildly different planetary satellites can be. Some very interesting moon-worlds could result in other systems. The variety of plausible planet-size satellites that may be found is immense. Some even provide almost all the familiar comforts of home.

The habitability of planets is also strongly affected by their locations and motions. Most of the planets in the Solar System pursue orbits that are reasonably close to circular. But this would not necessarily be the case for another stellar system. The orbital inclinations of the planets in the Solar System are low, except at the fringes. We have only one planet, Pluto, with a highly eccentric orbit. Some particularly badly behaved planets in other stellar systems may, like Pluto, range so widely that they actually cross the orbits of other planets. Also, because giant impacts are probably not rare, many planets, like Uranus, may have extreme axial tilts, or even, like Venus, rotate in the retrograde sense, opposite to their orbital motion. Rotation periods may be as short as a few hours, or very long. An occasional planet may, like Mercury, rotate in a harmonic relationship to its orbital period. Some, orbiting close to a star or a giant planet, may have been completely despun by tidal forces. Most of these problems are lethal or highly discouraging to life, and hopelessly difficult to fix.

Much of the evolutionary life of a planet depends on how hot it is, and on how much ultraviolet light it receives. This being so, what would happen to a planet orbiting a red dwarf? Or a hot, brilliant star that emits torrents of ultraviolet? What about all those giant stars? And by the way, aren't most stars double? What kinds of orbits are possible in mechanically complex multiple-star systems? Are there safe locations for life-bearing planets amid the whirling gears? Clearly, some neighborhoods are best avoided!

But there is more to a planet's setting than just the location of the one or two nearest stars: some stellar environments are incredibly densely populated. In the cramped quarters of globular clusters, for example, stars frequently pass very close to one another. During these close encounters of the big kind, a planetary system orbiting an average star may be broiled by passing alien suns. From time to time, the gravitational disturbance of a passing star may strip off one or more planets and hurl them off on independent orbits. In the long run, such runaway planets must escape from the cluster to wander about the galaxy in high-speed orbits. If it's not one thing, it's another.

Escaped planets ejected by (or stripped from) globular clusters will usually range far out of the congested central plane of the Galaxy. They must also dash through that central plane at very high speed twice per orbit. Such exploits would be very hazardous to any life forms on their surfaces. On such a planet you could die of loneliness — or simply freeze solid.

Many other hazards lie in wait for the unwary planet. Giant collisions and falling moons can sterilize them completely. Exhibitionist performances by nearby stars can sear their surfaces. And one's own star can sometimes be unreliable. There are a lot of things to look out for before you sign a purchase agreement.

What, precisely, makes a world habitable? Answering this question for life as we know it is hard enough. But what about life as we don't know it? Are there alien species that abhor water and are allergic to carbon? What other chemistries of life seem plausible? And how many habitable planets should we expect to find in our Galaxy? How far away might the nearest one be? Answering these questions gets us into the numbers racket, and makes us think hard about what makes a planet habitable at all, what makes it tolerable to Earth-born life, what makes it attractive to humans, and how planetary environments may be engineered to make them less hostile.

Many questions remain to be answered about the nature of other planetary systems. How do we find out the answers? It's unlikely that we'll be radioed an encyclopedia — we are, for now, on our own. What can we find by observing from Earth? Could we find Earthlike planets around other stars? Could we send automated interstellar probes to look at the planets of other stars? Could human astronauts travel between the stars to see these new worlds firsthand? Could we emigrate to the stars? Can humans adapt to unfamiliar conditions to meet a new planet halfway?

These issues lend a new poignancy to the familiar question, "What kind of world do you want your grandchildren to live in?"

THE PLURALITY OF WORLDS

WHAT IS THE NATURE OF the Universe? Where do we come from? Why are we here? Where are we going? Where do we, as living intelligences, fit into creation? What are those countless heavenly bodies really like? What, if anything, are they good for? Are they habitable worlds? Do they, or did they ever, actually harbor life—even intelligent life?

These questions are now raised almost daily by a flood of news stories relating the discovery of planets orbiting about other stars, or reporting evidence of simple life forms in meteorites that come from Mars. Certainly, all these questions are timely. But, astonishingly, they are also among the most ancient questions that have intrigued mankind. They have occupied the best minds of philosophy, religion, and science since the dawn of history. Indeed, the evolving answers to these questions are themselves a documentation of how we have changed—and where we are going. The philosophers of the Greek golden age; the Hindu sages who wrote the Shrimad Bhagavatam Mahapurana and the early Jyotish literature; Moses, the Old Testament prophets, and the early church fathers; and countless folk traditions around the world give us windows into the ancient human soul, allowing us to see—at least to glimpse—the world as it seemed to our forebears. Today, as we stare in astonishment at the Hubble Space Telescope deep-field image of layer upon layer of galaxies retreating back into the deepest depths of time, and as we read almost weekly of the discovery of strange worlds orbiting other suns, we share the deep feelings of awe and mystery, of reverence before an unfathomable and endlessly diverse Universe, that are no less real than those felt by the ancient seers. Whether scientist or artist or mystic, we are stunned by the beauty and complexity of what we see. But whether we look with the eyes of science or religion, we see order and lawfulness underlying superficial complexity. There is no trait more human, or more ancient, than the desire to seek understanding—to bring order to our experience and express that order in universal laws.

QUESTIONS ABOUT the heavens were discussed by the earliest Greek philosophers, starting with Thales and Orpheus in the sixth century B.C. Their naked-eye Universe consisted of Sun and Moon and several small planets ("wanderers") that moved against the background of "fixed" stars. Both argued that the Moon was, in some general way, similar to Earth, and that Earth, Moon, planets, and the pinpoints of light in the sky were all alike in kind; Thales called them all "stars."

In the fifth century B.C., members of two very different philosophical schools became intrigued with these questions. The early atomists Democritus and Leucippus emphasized their belief that all creation must be made out of vast (in fact, probably infinite) numbers of a few basic kinds of elementary building blocks. These individual particles, or atoms, all obey the laws of nature, which are everywhere the same. The atomists were inclined to conclude that there must be other Earths, other life, even other people in space. But the Greek concept of space was very different from ours. For example, Anaximander, a student of Thales', saw the Solar System as fitting snugly inside a sphere decorated with countless tiny specks of light, the "vault of the heavens," centered on a cylindrical Earth. This *kosmos*, or "world," was in effect a capsule surrounding Earth. The fourth-century B.C. Greek atomist Epicurus imagined countless spherical (perfect) *kosmoi* packed like a barrel of bubbles, each independent and self-sufficient, having no interaction with any of the others. Thus the idea of an *aperoi kosmoi*, or "plurality of worlds," was inspired by philosophical principles, not observational evidence, and was in any case devoid of practical significance: other kosmoi must lie outside our Universe, and were not even in principle observable. We would say that they existed outside the scope of physics, that they were "meta-physical."

The Pythagorean mystery school, in which Philolaus was trained, argued from the fact that the Moon and Earth moved around a common center; this reciprocity in their motions bespoke a symmetry in their natures. The Moon must, according to the Pythagoreans, be another Earth.

Thus two worlds, even two inhabited worlds, might reside in the same kosmos. Such an interpretation conflicted irresolvably with the idea that the heavens are heaven, whose contents are spiritual and perfect in nature, not material and imperfect. This idea of the supposed heavenliness of the heavens was furthered by Plato's doctrine of the ideality of spheres — the heavens, when viewed as spiritual, ideal, or immaterial, could relate to Earth only in metaphor, not in physical reality.

The redoubtable Aristotle, in the third century B.C., argued in his *De Caelo* (*Of the Heavens*) that each of the four elements (earth, water, air, and fire) strove to attain its own natural place or plane in our perfect Platonic, and therefore spherical, kosmos. The natural place of earth (solid matter) is in the center. Therefore it is quite impossible to speak of other earthy bodies, as a sphere can have only one center. Aristotle revisited this issue years later in his *Metaphysics*, asserting that a postulated plurality of worlds would require a plurality of "First Movers," which he rejected as an obvious impossibility.

The Judeo-Christian scriptures have almost nothing explicit to say about other worlds. Even the planets clearly visible to the unaided eye in the night sky were ignored by the ancient Israelites. If they left no record of Mercury, Venus, Mars, Jupiter, and Saturn, why should we expect them to have discussed other, more remote, and quite unseen worlds? Paul's Epistle to the Hebrews, dating from the first century A.D., begins with the first and only biblical reference to a multitude of worlds: "God, who at sundry times and in divers manners spake in time past unto the fathers by the prophets, hath in these last days spoken unto us by his Son, whom he hath appointed heir of all things, by whom he made the worlds." But even here it is not clear what Paul means by *worlds* — whether he is referring to the visible bodies of the Solar System, or speaking in broader terms.

The first century A.D. Roman natural philosopher Lucretius, in his *De rerum natura* (*On the Nature of Things*), fully adopted the atomist position and argued for an infinity of worlds: "There is, as I have said, no end; this truth speaks for itself, and blazes forth from the very depths of creation. Neither can we, seeing space reaching in all directions, infinite and free, and seeds [atoms] infinite in number flying in eternal motion through the void, suppose that only this one Earth and sky of ours has been created, and that those teeming remote bodies of matter serve no other use but ours. Further, as this world was fashioned by Nature through the intrinsic motion, collision, and joining of seeds [atoms], thrown about at random without any design, at best accreting when thrown together by chance, so must all great things begin—Earth, and sea and sky and races of living creatures. There must, I repeat, be such accretions of matter elsewhere, clasped, like our Earth, in the embrace of infinite space." Lucretius, who denies intelligence and purpose in creation, and attributes all change to the mindless collisions of a myriad of atoms "without any design," nonetheless explicitly assumes that every material object in the infinite, unbounded Universe exists for a purpose, to fulfill some function or serve some use. Lucretius's "argument from utility," after passing through numerous incarnations, survives in our own day in the form of proposals to make practical use of solar power and the material resources of space. This theme is explored in an earlier book of mine, *Mining the Sky* (Addison-Wesley, 1996).

The third-century Neoplatonist philosopher Plotinus raised a fascinating question with regard to these putative other worlds: "Could God make something better than that which he actually has made?" If He is omnipotent, He clearly could. If so, did He? If He did not do so, why didn't He? Note the implication that an all-powerful God would be capable of creating a world better than Earth. Christians, driven by the logic of divine omnipotence, it would seem, must accept the plausibility of Lucretius's opinion that habitable worlds are possible and even abundant; that life is widespread in space; and that there exists a plurality of Earths, with all that that implies. Thus predictions based on the theories of Plotinus and Lucretius, which respectively attributed Creation to divine power and to blind, godless chance, seem shockingly similar.

There were many points of Christian doctrine that seemed to bear on the issue of a plurality of worlds. For example, theologians generally agree that Christ's atonement was intended as a lifting of the burden of sin from the posterity of Adam. But the weight of original sin, which the people of Earth inherited from Adam, never fell upon the residents of other worlds. Therefore Christ's atonement could not apply to them. Those putative beings might therefore be lower life-forms incapable of sin (animals), or intelligent beings that had not yet fallen, or intelligent beings that had incurred sin but had never received an atonement. The last case points to irredeemably lost souls in Satan's grasp. But the idea that Christ may have sacrificed himself more than once on other worlds was, and is still, found offensive by most Christians. To some extent, the situation could be saved by denying the simultaneous existence of multiple populated worlds. Plotinus's contemporary Origen offered another possibility: that multiple peopled creations were acceptable so long as they were sequential, rather than parallel.

After the third century A.D., Western thought was put on hold for nine centuries. The Greek classics were largely lost — in fact, apparently irretrievable — after the burning of the Library of Alexandria in A.D. 391 by order of the Christian emperor Theodosius, whose pious hope was to purge the world of heathen writings. But many of the Greek classics survived in Arabic for centuries, unsuspected in the West. In the seventh century, during the long sleep of European civilization, Muhammad commented that there were "many worlds with lands and seas, and with human inhabitants," an idea whose antecedents seem more Greek than Arab.

In the wake of the first Crusades, much Arabic literature became available to Western scholars. In A.D. 1170 Gerard of Cremona translated Aristotle's *De Caelo* from Arabic into Latin, inspiring a searching examination of Greek wisdom by the Church. A thirteenth-century sage, Albertus Magnus, recognized the compelling interest of the issues raised by the Greek philosophers: "Since one of the most wondrous and noble questions in nature is whether there is one world or many, a question that the human mind desires to understand *per se*, it seems desirable for us to inquire about it." That spirit of inquiry was to mark the coming age.

In A.D. 1266, Roger Bacon, in his *Opus maius*, like Michael Scot and the Bishop of Paris, William of Auvergne, rejected other worlds on ostensibly Platonic grounds, arguing from the supposed perfection of the sphere: "If there were another mundus [kosmos] it would be of spherical shape, like this one, and there cannot be any distance between them, because there would then be a vacant space without a body between them, which is false. Therefore they must touch; but they cannot touch each other except at one point (by the twelfth proposition of Euclid's *Elements*), as has already been shown for circles. Hence everywhere except at that point of contact there must be void between them." Thus the principle that "nature abhors a void" became an argument against the plurality of worlds.

A year later, in 1267, Thomas Aquinas wrote in his *Summa theologica* that "the only ones who can assert that many worlds exist are they who do not acknowledge any ordaining wisdom, but rather believe in chance, as did Democritus, who said that this world, along with an infinite number of other worlds, was made from a casual confluence of atoms." Thus Aquinas made the concept of a plurality of worlds out to be a pagan philosophy, incompatible with Christianity. Ironically, this brought his conclusions nicely into line with those of the pagan philosopher Aristotle. Aquinas's argument would be echoed four centuries later by no less than Sir Isaac Newton.

Aristotle's writings were widely read, especially in the many universities that had been founded in the thirteenth century. In response to the spread of Aristotelian philosophy, Etienne Tempier, the Bishop of Paris, issued a *Condemnation* in 1277, attacking 219 common beliefs current in the universities. One of the propositions he denounced was the assertion that "the First Cause cannot make multiple worlds." In effect, Tempier affirmed the principle of the omnipotence of God. Because of Tempier's remarks, the trustworthiness of the Aristotelian basis of many of the arguments held by academia came into serious question, leading to a veritable landslide of anti-Aristotelian philosophizing.

The fourteenth century saw an acceleration and polarization of the debate over the plurality of worlds. The English philosopher William of Ockham, famed as the man who honed Occam's razor, was obliged as part of his requirements for the master of theology degree at Oxford to write a commentary on the *Sentences* (c. 1150) of Peter Lombard, Bishop of Paris. Ockham took on Lombard's Distinction XLIIII, which dealt with the recurrent question of Plotinus and others as to whether God could make the world better than that which he had made. Ockham boldly argued a slightly different question, "Could God make a world that is better than this one?" Ockham used this opening as an opportunity to argue for the plurality of worlds. His forceful challenge to this and other points led to Oxford's refusal to grant him his degree, and later to his excommunication.

In 1350 Jean Buridan, the rector of the University of Paris, argued that God had the power to create other worlds, and that the laws of motion may be different on other worlds. Although this sounds to the modern ear like heretical science, Buridan's meaning was rather less sensational: he argued that the "law" that dense materials (the element earth) returns naturally toward the center of Earth when displaced does not apply to other planets, whose dense matter would return toward *their* centers if displaced. In other words, he accepted a multitude of gravitating, material bodies as plausible. His theory presents a qualitative appreciation of gravitation, but is devoid of predictive power.

In the mid-fifteenth century, the German cardinal Nicholas of Cusa proposed that the Universe was infinite and unbounded (and therefore without a defined center), in which every location was more or less on an equal footing. In such a Universe, there was no obvious objection to the presence of vast numbers of worlds, many like Earth and each an independent center of attraction. In Nicholas's view, each world could, like Earth, be inhabited: "Life, as it exists here on Earth in the form of men, animals, and plants, is to be found . . . in a higher form in the solar and stellar regions." This seems to be a case of confusing "the heavens" with "Heaven."

The crystallization of the heliocentric theory of the Solar System by Copernicus in his *De revolutionibus orbium coelestium* (*On the Revolutions of Heavenly Spheres*) (1543) explicitly and successfully treated all the planets alike, drawing no distinction between the nature and location of Earth and those of the other members of the Solar System. This theory was criticized strongly by both Roman Catholic and Protestant clerics. Martin Luther himself fumed, "People give ear to an upstart astronomer who tries to show that the Earth revolves, not the Sun and the Moon. This fool wishes to reverse the entire science of astronomy."

One of the most remarkable characters in this drama, the exuberant hermetic mystic Giordano Bruno, heartily endorsed the Copernican system in his *La Cena de la Ceneri* (*The Ash Wednesday Dinner*) in 1584. Arguing from a principle of unity, and heavily influenced by Lucretius, Bruno claimed that all the glowing, fiery bodies of the heavens were stars, analogous with our Sun, each of which could rightly serve as a center for a system of planets. Bruno, on one of his many extended journeys to evade the Inquisition, brought the doctrine of the plurality of worlds to England and established it there, where it was to flourish for centuries to come.

Tycho Brahe, unconvinced by the Copernican system (he held out for the Sun as the center of motion of the other planets, which then circled Earth), remained resolved about the uniqueness of Earth. The absence of any detectable shifts in the positions of the stars as Earth orbited the Sun suggested to Brahe that they were huge bodies at enormous distances. He envisioned a vast gulf between Saturn and the stars, seeing these stars as so huge and hot as to be uninhabitable. From his argument that almost all of creation was a sterile wasteland, Brahe concluded that the idea of other worlds was false.

In 1605, Johannes Kepler expressed his opinion that "the Moon is correctly called by Plutarch a body such as Earth, uneven and mountainous, with even more mountains in proportion to its size than Earth," and suggested the moon's habitability. Kepler followed Brahe's argument from utility to a point, but reached the conclusion that those vast and distant suns could not be useless, and therefore must be attended by populated planets. But, curiously, Kepler's most powerful influence on this debate arose from his demonstration that the motions of the planets could be most easily explained by having them all pursue elliptical orbits around the Sun. Why was this so important? Because it called attention to the critical role of *observation* as the inspiration for and ultimate test of theory. In the dawning age of science, precise quantitative tests of theories quickly showed the inadequacy of the medieval penchant for reasoning by qualitative analogy.

In 1616, Tomasso Campanella, in his *Apologia pro Galileo*, addressed two of the most serious and damning attacks on the Italian astronomer Galileo Galilei. First, Galileo's assertion that there was water on the Moon and planets had been attacked on the Platonic and Aristotelian grounds that heavenly bodies were perfect, unchanging, and incorruptible. Allowing seas and mountains on the Moon "vilifies immeasurably the homes of the angels, and lessens our hope regarding heaven." In response, Campanella cited Genesis and Psalms regarding the "waters above the Earth" to dismiss this attack as an example of doctrinal impurity, neglecting Scripture in favor of slavish devotion to the opinions of the heathen Aristotle.

Secondly, it had been alleged by Galileo's critics that "if the four elements which form our world exist in the stars, it follows from the doctrine of Galileo that, as Mohammed declared, there are many worlds with lands and seas, and with human inhabitants. However, Scripture speaks of only one world and of one created man [but see Hebrews 1:2 and I Corinthians 15:45—49] so that this belief is opposed to Scripture." Campanella countered by arguing that having many small systems within a great Universe created by God was in no way a contradiction of scripture, only of Aristotle.

In 1632 Galileo, in his *Dialogue on the Two Chief World Systems*, argued that neither rain nor liquid water could exist on the Moon; nonetheless, life of a kind vastly different from that on Earth was still possible, adapted to local conditions. This appears to be the first clear statement of the possibility of truly alien life.

The great essayist and chronic insomniac Robert Burton wrote an admirably concise summary of this debate in his *Anatomy of Melancholy* (sixth edition, 1651): "We may likewise insert, with Campanella and Brunus, that which Pythagoras, Aristarchus Samius, Heraclitus, Epicurus, Melissus, Democritus, Leucippus, maintained in their ages, there be infinite Worlds, and infinite Earths or systems, in infinite aether, which Eusebius collects out of their tenents, because infinite stars and planets like unto this of ours, which some stick not still to maintain and publickly defend: I look for innumerable worlds wandering in eternity."

Elsewhere, Burton writes, "Kepler (I confess) will by no means admit of Brunus' infinite worlds, or that the fixed stars should be so many Suns, with their compassing Planets, yet the said Kepler, betwixt jest and earnest in his *Perspectives, Lunar Geography,* and his *Dream,* besides his *Dissertation with the Sidereal Messenger,* seems in part to agree with this, and partly to contradict. For the planets, he yields them to be inhabited, he doubts of the Stars: and so doth Tycho in his *Astronomical Epistles,* out of a consideration of their vastity and greatness, break into some such like speeches, that he will never believe those great and huge bodies were made to no other use than this that we perceive, to illuminate the earth, a point insensible, in respect of the whole. But who shall dwell in these vast bodies, Earths, Worlds, if they be inhabited? rational creatures? as Kepler demands, or have they souls to be saved? or do they inhabit a better part of the World than we do? Are we or they the Lords of the World?"

Many writers have offered richly contradictory answers to these questions. In a memorable example, Voltaire's *Candide* (1759) has the character Pangloss insist, with inane optimism, that this is indeed "the best of all possible worlds." We may view this as an irreverent, sarcastic, and belated answer to Plotinus.

Bruno received further English support for the idea of a plurality of worlds in 1638, when John Wilkins's *Discovery of a New World in the Moone* advocated the habitability of the Moon. The absence of references to other worlds in the Bible was, according to Wilkins, no more significant than the Good Book's failure to mention the known planets.

The French philosopher René Descartes's *Principia philosophiae* (1644) developed an elaborate pre-Newtonian theory of *tourbillons,* or "whirlpools," of material space-stuff. These whirlpools filled all space, butting up against each other with no intervening void. (This model ignores Bacon's objection from basic Euclidean geometry that any round structure cannot be stacked without leaving prohibited voids.) The static, eternal, unchanging kosmoi of Epicurus here become dynamic, evolving entities. Planets were in fact stationary in the invisible, incompressible medium in which they were embedded, transported by the whirlpool motion around their primaries. All stars, more or less, had systems, but these systems were by no means identical. The boundaries between whirlpools could, according to Descartes, be crossed by material objects such as comets. Logically, however, this theory is incapable of explaining the motion of comets, or of anything in an eccentric orbit, or of any bodies in intersecting orbits, since each of these violates the assumption of the embedding of planets in a whirlpool of space.

Descartes's entire basis for understanding the dynamics of stars and planets was overthrown by Newton in 1687. The idea of whirlpools of space, never more than an overgrown metaphor, was discarded in favor of the quantitative idea of gravitational force. The planets were maintained in motion not by a force pushing them, Newton posited, but by the absence of frictional dissipation of their energy of motion, which they had inherited at the time of their formation. The simple postulate of a gravitational force, dropping off as the square of the distance, was shown by Newton to give rise to Kepler's laws of motion. Newton, however, was very cautious in applying his theories to other worlds. His scientific speculations were always tempered by his theological beliefs.

The astronomer Thomas Wright, in his *Original Theory or New Hypothesis of the Universe* (1750), while sharing Newton's concern for theological correctness, set himself the task of placing the plurality of worlds on a firm Newtonian footing. He placed great emphasis on proving, both from observation and theory, that other stars are suns. Although he argued on observational grounds that the other planets of the Solar System were grossly similar to Earth, he based his arguments for planetary systems orbiting other stars squarely upon the principle of utility.

Immanuel Kant similarly argued, in his *Allgemeine Naturgeschichte und Theorie des Himmels* (*General Natural History and Theory of the Heavens*, 1755), that the material Universe is governed by natural law, and that that lawfulness constitutes direct evidence of the wisdom and power of God. Kant claimed that Newtonian forces could bring order to a chaotic, atomistic Universe and bring about the origin of stars and planetary systems. He argued by analogy with our Solar System that "these systems were formed and made in the same way as our own, out of the elementary atoms of matter that pervaded the void." He further argued that the Milky Way was a vast assemblage of stellar systems, and that other faint nebulae were in fact also such systems (we now call them galaxies). Kant links the development of intelligence and spirituality to the "quality" of the matter in which they develop. The densest regions he associates with minimal rationality, with successive levels of higher spiritual and rational attainment possible toward the more rarefied periphery. This theory was recently put to striking use in the science fiction novel *A Fire Upon the Deep* by Vernor Vinge.

By 1808 the French astronomer Jérôme de Lalande could casually dismiss the Cartesian model and endorse plurality without whirlpools, basing his arguments, like those of Kant, on the Newtonian theory of universal gravitation. The transplantation of Descartes's vision of infinite spaces and numberless worlds into the Newtonian Universe was by then an unqualified success.

The early nineteenth century was a time of progress on the observational front. The widely reported 1834–1838 South African expedition of John Herschel to survey the southern skies occasioned a great public furor when Richard Locke, a writer for the *New York Sun*, authored a series of stories purporting to reveal exciting results from the expedition. The stories built up in great detail, and with an occasional hint of plausibility, a wholly fictitious account of Herschel's observations of the Moon, which culminated in a revelation of vast forests, large animals, intelligent beings, and cities on the lunar orb. Even competing newspapers nodded in solemn approbation. Public opinion was prepared to accept life almost anywhere.

The first American edition of William Whewell's *The Plurality of Worlds* (1854) generally doubted the existence of life off Earth, but allowed that: "The surface of the Moon, or of Jupiter, or of Saturn, even if well peopled, may be peopled only with tribes as barbarous and ignorant as Tartars, Esquimaux, or Australians." Whewell approvingly cites the opinion of the German mathematician-astronomer Friedrich Bessel, who wrote in his *Populäre Vorlesungen Über Wissenschaftliche Gegenstände* (*Popular Lectures on Scientific Subjects*, 1848) that "those who imagine inhabitants in the Moon and Planets suppose them, in spite of all their protestations, as like to men as one egg to another." But Whewell, who wrote anonymously because he apparently anticipated opposition to his negative assessment of life on other worlds, concludes that he would not be surprised if humans were the only intelligent, moral race in the entire Galaxy. So positive was the prevailing American attitude on life in space, however, that a special introduction, written by the president of Amherst College, Edward Hitchcock, was added to the American edition of Whewell's text to commend the author's attempts to reconcile religion and science and to argue for a more liberal interpretation of the astronomical evidence. Curiously, Hitchcock, the nineteenth-century scientifically literate Protestant theologian and educator, asks the ancient question, "Of what possible *use* to man are those numberless worlds visible only through the most powerful telescopes?" His answer, echoing that of so many of his forebears, is that they are there for the benefit of other intelligent species: "When we see how vast is the variety of organic beings on this globe, and how manifold the conditions of their existence; how exactly adapted they are to the solid, the liquid, and the gaseous states of matter, can we doubt that rational and intelligent beings may be adapted to physical conditions in other worlds widely diverse from those on this globe? May not spirits be connected with bodies much heavier, or much lighter, than on Earth; nay, with mere tenuous ether; and those bodies, perhaps, be better adapted to the play of intellect than ours; and be unaffected by temperatures which, on Earth, would be fatal?" A daring conjecture indeed; that some intelligent beings may live free in "tenuous ether," without need of a planet on which to dwell!

The same general attitude was favored by the great French popularizer of astronomy, Camille Flammarion. Flammarion summarized the arguments for countless alien worlds in his book, *La Pluralité des Mondes Habitées* (*The Plurality of Inhabited Worlds*), (1889), a book written in 1861 but unable to attract a publisher until Mars became a hot topic due to Schiaparelli's famous "discovery" of canals on the planet during the 1877, 1879, and 1881 close oppositions of Mars. (Oppositions occur when Mars and the Sun are opposite in the sky; i.e., when Earth is passing close to Mars between it and the Sun.) The American astronomer Percival Lowell built an observatory near Flagstaff, Arizona, dedicated to the study of Mars and jumped on the canal bandwagon with an enormously influential book, *Mars*, published in 1895. In response, Flammarion's book was promptly translated into German, English, Spanish, Italian, Portuguese, Russian, Danish, Swedish, Polish, Czech,

Arabic, Turkish, Chinese, and braille. Flammarion personally presented an autographed copy of the book to Napoleon III. The ideas in it were not only widely familiar to experts; they were widely discussed by the public. Flammarion's central theme was "Il approprie a chaque astre une humanité." (A human race takes over at every star.) Victor Hugo, upon reading *Pluralité*, wrote to Flammarion, "Je pense comme vous." (I think as you do.)

Thus, by stages, the idea of a cozy, Earth-centered kosmos, in which Earth was the sole physical body, yielded to the conception of a full-blown stellar Universe of countless stars, each the center of a planetary system of its own, with life and intelligence widely distributed through them. Worlds unlike Earth were conceived as able to bear their own indigenous forms of life, well adapted by origin and habituation to a wide range of conditions, many of them perhaps un-Earthly in degree, or even in kind. The question whether we, or they, were the true lords of creation remained unanswered. Flammarion ushers in the familiar twentieth-century Universe within which space travel and contact with alien races become at least conceivable, and perhaps inevitable.

THE LONG human tradition of imaginative voyages, when fertilized by the dawning acceptance of the plurality of worlds, gave rise to both literary and scientific interest in these new possible arenas of activity. Through astronomy, the planets become not "immaterial" or "celestial" lights in the sky, or mythological figures, or divine portents, but other places, subject not only to study and theory but even to fictionalization. In the light of the age of exploration, it was at least conceivable that one could visit other places and explore them; and it remained only for chemistry, physics, and engineering to make the conceivable practical. It is from such a wedding of vision and science that the space age, the era of human exploration of other worlds, has sprung.

Human imagination long preceded human technology in probing the mysteries of space. The first tale of travel to another world, by Lucian of Samosata in the first century A.D., invokes a waterspout to carry its hero into space. There he finds the Sun and Moon not only populated, but their denizens at war with each other over colonial rights on Jupiter. The whole piece is an airy fantasy, lacking even such discipline as might have been provided by heeding the rudimentary science of Aristotle.

The popular imagination was also stirred by fabulous tales of journeys to foreign and exotic lands. Marco Polo's thirteenth-century accounts of his exploits on the road to China and in the court of the Great Khan, although generally factual, were received as fantasy and mocked by many in Europe. John Mandeville's bizarre tales of his fourteenth-century travels in the land of Prester John seemed to confirm to many readers the idea that travel stories were nothing but rank fantasies, even though many credulous souls read them as factual accounts. Both those predisposed to accept these tales of fabulous travels and those inclined toward skepticism would find ample stimulus in the next round of imaginative innovation, the extension of these voyages to the Moon and beyond.

Marco Polo brought back from the East knowledge of gunpowder and of small rockets powered by special formulations of gunpowder. These small solid-propellant rockets were suited for light military use, or for launching fireworks for entertainment. Use of these rockets spread rapidly from east Asia through south Asia to Europe. But gunpowder rockets are teacherous and unsuited for scaling up to sizes more than a few feet long. The culmination of this technology was the Congreve rocket used by England during the War of 1812, by whose "red glare" the American national anthem was penned. These rockets were modest improvements over the Chinese and Persian war rockets of six centuries earlier.

The seventeenth century saw not only the introduction of the astronomical telescope and the wide publication of its discoveries, but also a liberation of the imagination. Soon after Galileo reported mountains on the Moon, revealing the lunar orb to be a physical, imperfect body rather than an immaterial, perfect sphere, Kepler's *Somnium seu astronomia lunar* (*Dream*) told of an imaginary trip to the Moon. As the title suggests, no attempt was made in the narrative to provide a viable means of travel through space. The purpose of the piece was simply didactic; to instruct the reader in the lore of the Moon without seeming overly academic. But, in 1634, Kepler did foresee and affirm the possible physical exploration of space, even though the means of such travel were beyond the scope of his science: "When ships to sail the void between the stars have been invented, there will be men who come forward to sail those ships."

The year 1638 saw the publication of Bishop Francis Godwin's *The Man in the Moone or a Discourse of a Voyage Thither by Domingo Gonsales.* In this story, the shipwrecked Gonsales is transported by means of a flock of trained swans which, unexpectedly, migrate to the Moon.

Equally fantastically, in Cyrano de Bergerac's *Voyage dans la Lune* (*Trip to the Moon*, 1657) the protagonist is carried aloft by the power of the Sun's attraction on vials of dew.

Margaret Cavendish, an amateur scientist of some note, wrote in *The Description of a New World, Called the Blazing World* (1668) of a fictional visit to a world, unaccountably invisible, which is in contact with Earth at the North Pole. The hapless visitors are blown there by a freak storm. Cavendish characterized this work as "meerly Fancy," perhaps an apt description of a world abounding in both reason and wealth.

Daniel Defoe's 1705 fantasy novel, *The Consolidator: or Memoirs of Sundry Transactions from the World in the Moon*, was a true "fantastic voyage," really no more than an elaborate political satire, purporting to be set on the Moon as a means of piquing interest and disavowing realism. It is plausible that Defoe was encouraged to experiment with such a theme by the success of Margaret Cavendish's novel a generation earlier. Jonathan Swift's *Travels into Several Remote Nations of the World* (1726), reprinted as *Gulliver's Travels*, was yet another of these fantastic voyages, clearly inspired in its turn by Defoe's *Robinson Crusoe*. Swift confined Gulliver's wanderings, like those of John Mandeville, to Earth; but again, like Defoe's *Consolidator*, the principal purpose was political satire and social commentary.

The first seed of a physically plausible means of spaceflight was planted obscurely in the ordinary book of the polymath Erasmus Darwin, grandfather of Charles Darwin. In these notes, dating from about 1779, Darwin's biographer Desmond King-Hele has found an evocative sketch of a simple liquid-fuel rocket engine, with hydrogen and oxygen tanks connected by plumbing and pumps to an elongated combustion chamber and expansion nozzle, a concept not to be seen again until the 1890s. The design also features the ability to use air as the oxidizer, making it technically a ramjet/rocket hybrid, an engine type that would eventually attract attention again in the 1990s. The design, like those in Leonardo da Vinci's notebooks, lay unrecognized for two centuries. Since Darwin was aware of the diminution of air pressure at high altitudes, the inclusion of an oxygen tank makes it clear that he was thinking of, at the least, very high-altitude flight. But Darwin provided no explanatory or descriptive text, and the idea was lost to posterity until King-Hele expounded on the diagram.

It was not until 1865 that Jules Verne rendered into prose the first physically-based space flight, in *De la Terre à la Lune* (*From the Earth to the Moon*). Verne employed a titanic gun, based loosely on Civil War technology, to fire a three-man sealed capsule to the Moon from a Florida launch site. There are numerous physical flaws in the story: the acceleration achieved by such a launch vehicle would have utterly crushed the hapless passengers, and Verne erroneously has them experience weightlessness only at the point of equal gravitational force between the Moon and Earth.

The famous American writer and orator Edward Everett Hale wrote of the launching of an artificial Earth satellite in his story "The Brick Moon," published in the *Atlantic Monthly* in 1869. Hale invokes centrifugal force tapped from two gigantic counter-rotating flywheels (ultimately powered by a waterfall in Maine) to launch his masonry sputnik. It is unclear whether he had a well-defined (but mistaken) idea about the physics of acceleration, or whether he simply invoked scientific-sounding terminology to obscure the absence of an idea. (It was this same Hale who delivered the eloquent hours-long declamation at the dedication of the Gettysburg battlefield, only to be upstaged by Lincoln's polished little gem of a speech.)

In 1870 Jules Verne published *Autour de la lune* (*Around the Moon*), the sequel to *De la Terre à la lune (From the Earth to the Moon.)* The interest in this book comes mainly from the fact that the author invokes a "midcourse correction" while flying near the Moon, an idea ridiculed in a later editorial, written to attack Robert Goddard, in the *New York Times* of January 13, 1920. The author of the editorial pontificated with glorious arrogance and total lack of understanding of physics, "That Professor Goddard, with his 'chair' in Clark College and the countenancing of the Smithsonian Institution, does not know the relation of action to reaction, and of the need to have something better than a vacuum against which to react—to say that would be absurd. Of course he only seems to lack the knowledge ladled out daily in high schools." Although unacceptable to the opinion shapers at the *Times*, Verne's technique, embraced by Goddard, was actually used by the Apollo missions to the Moon. (Nowadays, high schools daily ladle out the knowledge that the exhaust gases of a rocket and the rocket's thrust chamber push against each other, satisfying Newton's laws while working admirably in a vacuum.) Verne's astronauts, peering at the Moon through a glass, also debate endlessly about the existence of forests, seas, and cities on the lunar surface.

Percy Greg's 1880 novel, *Across the Zodiac*, features an antigravity-powered ship complete with a hydroponic air-regeneration system, in which the hero visits a utopian civilization on Mars. The device of antigravity is also invoked in the 1894 novel, *A Journey in Other Worlds*, by Harvard graduate J. J. Astor to escape from the humdrum existence on a utopian Earth in the year 2000 by visiting Jupiter and Saturn. On Jupiter the novel's protagonist finds a sort of Jurassic dinosaur playground; on Saturn, a civilization of spirits that teach biblical principles. Later in his career, John Jacob Astor IV was to build the Waldorf-Astoria Hotel in New York before going down with the *Titanic*.

The German physicist Hermann Ganswindt, as early as 1891, was lecturing on the physics of reaction engines and advocating space travel using rocket propulsion. Unfortunately, it appears that he left no published record of his ideas. The Russian engineer A. P. Fedorov wrote a narrative titled *A New Principle of Aeronautics* (1896) on spaceflight.

Kurd Lasswitz's 1897 science fiction novel *Auf Zwei Planeten* (*On Two Planets*) describes a Martian invasion of Earth. The aliens quickly sink the British Navy and establish a utopian world government. The connection between these two events was presumably evident to Lasswitz's German audience. Sensing a potential English market for such fiction, in 1898 the British novelist Herbert George Wells wrote *The War of the Worlds*, in which the Martians, shot in giant artillery shells from Martian cannons, land in England. The alien invaders, with "intellects vast and cool and unsympathetic," are militarily unstoppable, but fortunately succumb to terrestrial microbes. American audiences are more familiar with Orson Welles's radio version, which was set largely in New Jersey, and a printed adaptation that recasts the novel in the geographical context of the Boston area.

The year 1898 also saw the premiere of Edmond Rostand's classic play, *Cyrano de Bergerac*. In one scene the redoubtable Cyrano, trying to divert the attention of the Comte de Guiche, pretends to have just fallen to Earth from the Moon, and promises to describe six different methods of interplanetary travel.

At the very time that Rostand was writing, the young Russian genius Konstantin Edouardovich Tsiolkovskii was already hard at work on the theory of rocket propulsion. In 1903 Tsiolkovskii published his treatise, *Exploration of Cosmic Space with Reaction Engines*, in which the theory of rocket propulsion was developed for the first time. This visionary work was unfortunately available only in Russian. Among its limited readership, it seems to have impressed only one prominent scientist, the eminent chemist Dmitri Mendeleyev.

Contemporary fiction was still stuck in the previous millennium. Edwin Lester Arnold's 1905 "scientific romance," *Lieut. Gullivar Jones: His Vacation,* has his protagonist make a wish and travel to Mars on a flying carpet. In 1917 the famed author of the Tarzan books, Edgar Rice Burroughs, published *A Princess of Mars,* in which his protagonist, John Carter, with similar technical sophistication, is drawn "with the suddenness of thought" to Mars. This nonphysical approach to interplanetary travel seems to have reached the end of the road in 1922, when E. R. Eddison has the hero of his fantasy novel, *The Worm Ouroboros,* travel to Mercury in a dream (the character *thinks* it is a dream—he can pass his hand through any material object he sees—but a Mercurian insists on his own authority that it is not a dream).

As the fantasy of space travel faded away like the dream it was, the dawning technology of rocket propulsion advanced energetically. From 1905 through 1907 the Norwegian astronomer Sverre Birkeland carried out experiments with small hydrogen-oxygen rocket engines, of which there are no known surviving records. In 1912 the French inventor Robert Esnault-Pelterie held a conference on rocket propulsion in Paris. He concluded that rocket flight into space would not be possible until nuclear energy could be mastered.

Having failed to make much of an impression on the world with his technical publications, in 1913 Tsiolkovskii published a science fiction novel, *In the Year 2000* (*On the Rocket*), about a group of rich men who build a rocket powered by hydrocarbons and liquid oxygen. These pre-Bolshevik investor-inventors can best be described as venture capitalists. One can only wonder how history might have been changed had Tsiolkovskii somehow come to the attention of his contemporary John Jacob Astor IV. Also in 1912–1913, Robert Goddard, having received his doctorate at Clark University, spent a fruitful year at Princeton University's Palmer Physical Laboratories, researching rocketry and pursuing basic patents in his spare time. While there he had many interactions with the great astronomer Henry Norris Russell, who advised him on his rocket patents.

In 1915 the Transylvanian inventor Hermann Julius Oberth submitted his designs for war rockets to the Austro-Hungarian Ministry of War, which invited him to abandon "this fantasy." Only months later, Professor G. Tickov of Pulkovo gave a public lecture on the importance of rocketry, in which he claimed that "rocket propulsion will remain necessary until gravitation can be annulled — probably by electrical means." As of this time, Tsiolkovskii, Oberth, and Goddard were all completely unaware of each other's existence.

Goddard, in 1918, wrote an astonishing futuristic essay, "The Ultimate Migration," in which he predicted the eventual flight of mankind out of the Solar System in hollowed-out asteroids to escape the death-throes of the Sun. But Goddard was anxious lest his practical efforts to build a high-altitude rocket be discredited by association with such grandiose schemes. So concerned was Goddard with professional credibility and public perception that he sealed up the unpublished manuscript in an envelope labeled "Special Formulas for Silvering Mirrors" and set it aside. It was finally opened over fifty years later, long after Goddard's death. The infamous 1920 editorial mentioned earlier clearly demonstrates that Goddard was correct in his perceptions.

The climate of opinion was no kinder to young Oberth. His doctoral dissertation on the physics of rocket propulsion and space travel was rejected by the University of Heidelberg in 1922, and remained unpublished until Oberth issued it at his own expense in 1929, under the title *Wege zur Raumschiffahrt* (*Routes to Space Travel*). But in the meantime, Goddard had built, tested, and flown liquid-fuel rockets. The space age was dawning. Ironically Oberth, unaware of Goddard's essay "The Ultimate Migration," often criticized Goddard as an unimaginative plodder who could not see as far as space.

In the same year, British physicist J. D. Bernal wrote in his essay "The World, the Flesh, and the Devil," of solar sails, beamed microwave power, solar power satellites, and the use of space resources. He proposed, quite independently of Goddard, the idea of hollowing out ten-mile asteroids for use as interplanetary habitats and interstellar "slowboats."

The idea of rocket travel was suddenly everywhere, its spores spread by pulp magazines. Hugo Gernsback's *Amazing Stories* carried, beginning in August 1928, E. E. "Doc" Smith's "The Skylark of Space," in which the hero flits about space in an antigravity ship, visiting not only the Solar System but also other stars. In his mind-blowing 1931 novel *Last and First Men*, British writer Olaf Stapledon casually assumed the use of rockets to settle Venus and Neptune. Rockets and space travel had entered public awareness. It was only a matter of time before rocketry would give us first-hand access to the planets and satellites of the Solar System, letting us for the first time appreciate the enormous diversity of worlds.

In the decade from 1930 to 1939 the German Verein für Raumschiffahrt (Union for Spaceflight), the British Interplanetary Society, the American Interplanetary Society (later the American Rocket Society), and the Russian GIRD were founded to investigate and advocate rocket propulsion and interplanetary travel. And Goddard's rockets were flying to show the way.

One of the proponents of the practical benefits of space was a young British engineer who suggested as early as 1939 in the *Journal of the British Interplanetary Society* that manned missions to the Mars system might derive great benefit from extracting water and other volatiles from the Martian moons Phobos and Deimos to refuel their rockets for descent to Mars or return to Earth. A few years later he proposed that satellites placed in high orbits above Earth's equator could serve as communications relays covering nearly the entire planet. That young man, Arthur C. Clarke, is today best known for his *2001: A Space Odyssey* and numerous other science fiction stories and novels.

From the 1930s through the 1960s, practical engineers such as Wernher von Braun in Nazi Germany and Sergei Korolyev in the Soviet Union brought the liquid-fuel rocket to a high level of performance and reliability. When von Braun was being interrogated by American officers at the end of the Second World War, he was astonished by their questions probing into the "secrets" of Nazi rocketry. He answered in apparent sincerity that the theory had been worked out by the American Goddard, and German engineers were just following his lead. The American officers were equally astonished. They had never heard of Goddard. But, accelerated and encouraged by the performance of the German V2 rocket, the world hastened on into space with considerable confidence that the goal could at last be realized. From the launching of *Sputnik 1* in 1957 to the first Apollo landing on the Moon in 1969, American and Russian scientists and engineers fairly leaped into the space age.

By 1974 Princeton University physicist Gerard K. O'Neill had proposed permanent large space colonies capitalized by exporting power to Earth. In 1977, at the instigation of physicist Peter Glaser of Arthur D. Little, Inc., and O'Neill, a NASA study of solar power satellites was carried out, and another NASA study in 1996 explored more advanced technical

options that promise to make SPS systems more economical. Princeton's Space Studies Institute, founded by O'Neill, continues to advance both the study of economic development of space and public awareness of its potential.

THE EDIFICE of our insights into other worlds rests in part on two distinct foundations: the "imaginary voyages" of Verne, Wells, Stapledon, and their fellow writers of fiction, and the "visionary technologies" of Tsiolkovskii, Goddard, Oberth, and their fellow scientists and engineers who have carried us to other worlds. It should come as no surprise to hear that the practitioners of these two fields inspired and encouraged each other. Goddard acknowledged his inspiration from Wells, and Tsiolkovskii spoke of the formative influence of Verne on his work. Gernsback tried to commission Goddard to write articles for his magazines. Likewise, the leading writers of science fiction built firmly upon the best available scientific data and technology: the rocket moved quickly from Goddard's test stand to the pages of Stapledon and others, and Wells hastened to build on the latest Mars observations and conjectures of Percival Lowell.

The space age made possible by these visionaries has given us true, firsthand access to our Solar System. We are no longer dependent upon astronomical observations, made through Earth's cloudy, dusty, hazy, and turbulent atmosphere, of planets that are mere dots in the sky. Since 1960, spacecraft built on Earth have been heading out to visit the Moon, Mars, Venus, and even Mercury, Jupiter, Saturn, Uranus, and Neptune. The first tentative missions to comets and asteroids have already returned priceless data. Vicariously, through the eyes and hands and sensors of these robot craft, we have begun to visit alien worlds and immerse ourselves in their mysteries.

In a way, the recent discovery of planets orbiting stars transports us back to where we were in the 1920s in our study of the Solar System. In fact, the situation is even worse: although our instruments are vastly more sensitive and capable than those we had then, we are charged with the task of observing planets that are roughly a million times as far away as our neighbors in the Solar System. In the usual sense of the word, we have not yet "seen" these planets at any wavelength, but simply detected the gravitational effects of their presence on the motions of their parent stars. Even the crudest images of these bodies do not yet exist. We can, from the observations that are available, often deduce their orbital periods, eccentricities, masses, and distance from their stars, and hence estimate their surface temperatures, diameters, escape velocities, and gravities. But as long as we cannot "see" them in any way, our conclusions must remain indirect and inferential.

Fortunately, we already know something about planets: our Solar System contains a group of bodies of very diverse size, distance, temperature, composition, and structure. We see enough diversity in our planetary neighbors to realize that these bodies are a sample—though incomplete—of all the possible kinds of planets that may exist. We are inspired to examine the countless types of alien worlds that would result from only modest changes in the origin and evolution of our own system. Our basis for understanding the worlds of other suns must be the planets we see nearby, the very ones that our spacecraft can fly by, orbit, land on, and explore. We therefore turn our attention to the Solar System before leaping into the interstellar void. We do so not because it holds all the keys to understanding the kosmos, but because it is convenient. Pragmatically, we start our search under the nearest streetlight—the Sun.

2

WORLDS AS WE KNOW THEM

NO NINETEENTH-CENTURY ENGLISH GENTLEMAN could count his education complete until he had undertaken a yearlong pilgrimage to the cultural centers of continental Europe. The purpose of this grand tour was to flesh out the dry book-learning doled out at dank and cloudy Oxbridge with firsthand experience of Paris, Rome, Venice, Florence, Athens, Vienna, and the like, seeing the Mona Lisa and the Pieta, the Pantheon and the Parthenon, soaking up the warm classical sun by day and the arias of Rossini, Donizetti, Bellini, Verdi, and Puccini at La Scala by night. Gray abstracted knowledge came alive and personal under flawless Wedgwood-blue skies. Nothing can replace actually *being there.*

The same principle has applied to twentieth-century humans. After centuries of viewing the planets through a telescope darkly, rocketry made it suddenly possible to encounter them face-to-face. The best telescopic views of Mars, viewed through our turbulent and cloudy atmosphere, showed a blurry disk with seasonal white polar caps and great irregular expanses of dark and bright reddish terrain. But, starting in 1960, spacecraft departed from Earth to fly by Mars and Venus at close range, and then to impact them, orbit them, and land gently on their surfaces. Cameras aboard these spacecraft transported us to new worlds, showing us in living color what it would be like to stand on their surfaces or orbit above their cloud tops. The Moon, first assailed by small automated man-made probes in 1959, bore human footprints within ten years.

By 1962, experts in celestial mechanics had pointed out that the late 1970s would present a rare planetary alignment, a sterling opportunity to send a spacecraft from Earth to Jupiter, Saturn, Uranus, and Neptune on a single arc. The Voyager spacecraft were built and launched to take advantage of this alignment. This mission, visiting five of the nine planets, soon became known as the Grand Tour. (The last previous Grand Tour opportunity was missed because space was not a high priority of President John Quincy Adams.)

Now, at the end of the twentieth century, we can look back on decades of space-based studies of the Solar System. Building upon our recent explorations, we can take a Grand Tour of the Imagination. We can interpret what we have learned of the distinctive individual traits of the planets to help us form generalizations about the nature of planets, their family relationships, and the trends in their properties discerned from our study of the Solar System. Such an overview provides the best available preparation for understanding the planetary systems of other stars.

THE SOLAR System contains a stunning diversity of bodies. The inner planets are rocky, meaning that they are made of mixtures of metallic, oxide, silicate, and sulfide minerals. But each large rocky body (Mercury, Venus, the Moon, Earth, Mars, Io, the largest asteroids) is unique in a variety of ways. The smallest rocky bodies have never melted. The largest, however, have melted and differentiated into layers of varying density and composition (core, mantle, crust, ocean, neutral atmosphere, ionosphere), crudely reflecting the "elements" delineated by the ancient Greeks: earth, water, air, and fire. (This point of view exposes the geocentric bias of the ancients: among all the known planets, the only one offering all these substances in abundance is Earth itself!) The other large satellites (Europa, Ganymede, Callisto, Titan, Triton) and the Kuiper belt* bodies and their escapees (Pluto, Charon, Chiron, and so on.) contain not only rock, but also large proportions of water ice and other frozen volatiles. These ice-plus-rock bodies constitute a continuous series ranging from purely rocky bodies to those with high content of volatile materials. Not surprisingly, all of these entities have distinct idiosyncrasies. Finally, the four giant planets are dominated by gases. All four are made mostly of hydrogen and helium,

*The Kuiper belt is a broad, diffuse belt of ice-rich bodies, like a giant asteroid belt, extending out to about 200 AU from the Sun. The largest of these bodies, Pluto, is 2304 kilometers in diameter, but only a few of the largest surpass about 700 km. The orbits of most icy asteroids have low inclinations, although many have rather high eccentricities, so that at perihelion they cross the orbits of one or more of the giant planets.

with varying amounts of the ice-forming and rock-forming elements. The largest of these bodies approaches the Sun in composition. Thus we may crudely separate the bodies in the Solar System into three categories: rock, dirty ice (a mixture of ice and rock), and solar. By solar we are referring to the composition of the Sun at the time of the origin of the Solar System, not to the modern Sun. It is important to remember, though, that the entire system, and all bodies in it, with all their stunning diversity, originated in a presolar cloud of gas and dust that had a composition very similar to that of the Sun.

More useful than categories of planets, however, is the division of the *materials* of the planets into the three major categories of rock, ice, and gas. Each of these categories includes a range of composition, depending on the circumstances under which the material was formed. Thus, for example, a portion of the presolar cloud that is just cold enough to condense water ice would be too warm for condensation of ammonia, methane, carbon monoxide, nitrogen, or the inert gases. Therefore water ice would condense relatively free of impurities and be available as a mineral for incorporation into newly forming solid planets. But at slightly lower temperatures, ammonia would also condense, and at even lower temperatures, methane would condense. The crucial factor determining what materials are condensed and available to be incorporated into growing solid planets is temperature.

The systematic trend that dominates in the Solar System is easily described: the most volatile materials are retained efficiently only at far distances from the Sun. This is not surprising, since we expect the temperature in the original presolar gas cloud (the Solar Nebula) to have been highest close to its center and lowest toward its periphery. Hydrogen and helium, the most volatile materials of all, apparently never condense: they are captured only by planets that are massive enough (and cold enough) that they can hold these light gases indefinitely in their gravitational fields. This suggests a simple explanation for the general trend in the composition of the planets with respect to their distance from the Sun: the composition of solids in the Solar Nebula reflected a general dropoff of temperatures the farther the distance from the center. Indeed, knowing the vapor pressures of these substances and the overall composition of the bodies in the Solar System, we should be able to reconstruct how the temperature in the Solar Nebula dropped off with distance from the center. In reality, many complicating factors enter in to make this a difficult task.

The two most important problems are that the composition of most of the planets is very imperfectly known, and that local environmental conditions, such as the temperature and the location of small bodies in the Solar Nebula, may have changed in a complex and largely unknown manner over time.

Suppose that temperatures in the Solar Nebula did drop off systematically the greater the distance from the center; the composition of the solid particles in the nebula must then have varied systematically depending on their distance from the Sun. A planet growing by accretion of small solid bodies at Earth's present location (1.000 AU from the Sun) would accumulate most of its material from a nearby ring of the nebula. Crudely speaking, solids that were closer to the growing Earth than to Venus or Mars would most likely end up as part of Earth. During the later stages of the sweeping-up process, the half-grown planets themselves would kick some of the solids around and cause some degree of radial mixing: some Mars-region material, and some from near Venus (and an even smaller amount from farther away), would find its way into Earth. Mars in turn would be made mostly of local material, formed farther from the Sun and at lower temperatures than Earth, but it also would contain material kicked out from Earth or expelled by Jupiter's gravitational influence from the asteroid belt. The fact that all of the terrestrial planets differ in composition simply means that they were assembled at different distances from the Sun, and sampled the same raw materials in very different ways.

In other words, if the planets had all started growing, say, 0.10 or 0.15 AU farther from the Sun, each would have incorporated the available raw materials differently from that of any planet in our Solar System. Each of these planets would end up with a composition that does not match any planet in the Solar System. A planet formed at 0.85 AU would be intermediate in average composition between Venus and Earth, even though the starting materials were exactly the same. Allowing also for random mixing events (the capture of a Moon-size piece of asteroid-belt material by Venus, for example), even a suite of chemically identical nebulae could give rise to a far wider variety of planets than we actually have. We must conclude that, although the composition of the planets in our Solar System is very diverse, they comprise a very incomplete picture of all the planets that could have been formed from the very same raw material. If we also allow for other stars having different compositions, then the solid materials in their nebular disks must have been different from those in ours; thus the planets formed there *must* be different in composition from any planet in our Solar System.

All this still does not take into consideration differences due to factors other than composition, such as variant planetary masses. A planet with the composition of Earth and the mass of the Moon would not be very interesting, because it would lose it oceans and atmosphere readily. Clearly, the planets of the Solar System as we know them offer a very limited sample of reality, and are an inadequate guide to the range of possible worlds. Worlds unknown vastly outnumber the worlds we know. As to whether our Earth is the best of all these possible worlds, we have far to go before we can even offer an informed opinion.

3

GENESIS AND EVOLUTION

WE CAN SAY NOTHING ABOUT the habitability of a planet unless the star it orbits is well understood. We must know how stars and planets come into existence, and how they change with time, before we can assess their suitability for the origin and development of life.

Two separate scientific constituencies are interested in the origin of the Solar System. The first of these is the community of astrophysicists whose concern is the origin of the Sun and stars, and who regard the planetary system as an unnecessary complication. The other is the community of planetary scientists, who care primarily about the origin of planets, moons, meteorites, asteroids, and comets, and who have traditionally shown disinterest in the stars. Each year, these two communities have come to pay ever more attention to each other's theoretical concerns and sources of data, all because of an astonishing fact: the scenarios developed by astrophysicists for the origin of stars and those developed by the planetary scientists for the origin of planetary systems have turned out to be shockingly similar. When conditions favor the origin of a star, they also favor the origin of planets.

Our Solar System has much to teach us about the genesis of stars and planets. Drawing from the survey of the Solar System in Chapter 2, we can identify a number of phenomena that any serious theory of Solar System origins must explain, such as the shapes of the orbits of the planets and their masses. For example, the planetary system is confined to a flattened disk out to distances of at least 100 AU from the Sun. The overwhelming majority of the mass of the planetary system orbits within about one degree of the common plane. Only a very few tiny bodies (a few dozen errant asteroids) occasionally rise far up out of the common plane.

Further, all these bodies orbit the Sun in the same direction, counterclockwise as viewed looking down on the north pole of the Solar System. The rotational directions of the planets and the Sun conform fairly well to the same rule, but here there are interesting deviations: Mercury is in its 3:2 spin-orbit resonance; Venus has a very slow retrograde (clockwise as seen from the north) rotation; Uranus's spin axis is almost in the plane of its orbit, so that it rotates "on its side"; and the Pluto-Charon system is strongly tipped. Also, Earth, Mars, and Saturn have axial tilts on the order of thirty degrees.

Every orbit is described by its shape, size, and orientation. The orbits of interest to us are stable, closed orbits in the shape of an ellipse. The size is given by the mean distance (also called the semimajor axis) of the body from the star (or planet) around which it revolves. The orientation of the orbit is given by angles describing its tilt and the direction in which the semimajor axis points.

The orbits of several of the planets are very nearly circular, but three of the planets have significant orbital eccentricities. The eccentricity is simply a measure of how far the shape of an orbit departs from a perfect Platonic circle: $e = 0$ is a circle, and $e = 1$ is an orbit so elongated that the orbital period is infinite; in other words, it is an ellipse stretched to the point that the second focus is at infinite distance from the prime focus, where the Sun is located. Such an orbit is better described as a parabola than an ellipse. Any eccentricity between 0 and 1 describes an elliptical orbit. For example, consider the eccentricities of the orbits of the three pairs of closely related planets in our system: the "little sister" planets Venus and Earth have eccentricities of 0.00678 and 0.01670, the "big brothers" Jupiter and Saturn have 0.0479 and 0.052, and the "middle-sized brothers" Uranus and Neptune have eccentricities of 0.050 and 0.004, all of which are quite low. The two planets at the boundaries of the Solar System, Mercury and Pluto, have orbital eccentricities of 0.2056 and 0.2566, much higher than the others. But Mars, the planet closest to the disturbing influences of giant Jupiter, also has a modestly elevated eccentricity, 0.0934.

The inclinations of the planetary orbits also carry evidence of how the planets formed. Taking Earth's orbit as the reference plane, all the other planets have orbits inclined less than 3.4 degrees — again, with the exception of Mercury (7.00 degrees) and Pluto (17.13 degrees) at the edges of the system.

The pattern of distribution of mass seen in the Solar System is also clear: the inner, rocky planets have masses ranging downward from one Earth mass, whereas the giant planets fall into two groups: Jupiter and Saturn (318 and 95 Earths, respectively), and Uranus and Neptune (14.6 and 17.2 Earths, respectively). The densities of the planets clearly show that the inner (terrestrial) planets are rocky, the Uranian pair are about half gas and half a rock-ice mixture, and the Jovian pair are close to the Sun in overall composition, with only a few percent of ices and rocks. The icy satellites, Pluto, and comet nuclei are surely ice-plus-rock mixtures; the Kuiper belt and Oort cloud bodies almost certainly fall into the same category.

The satellite systems of Jupiter, Saturn, and Uranus remind us of little solar systems: a number of nearby satellites in coplanar, low-eccentricity orbits, but sometimes with a distended cloud of outer satellites with much more chaotic orbits. In Jupiter's system, where we can discern compositional trends, we find a strong suggestion of rocky material close to the planet and rock-plus-ice mixtures farther out, emulating the composition trend we see in the Solar System itself.

All of these general conclusions reflect how the Solar System formed. They are all valuable clues to help us unravel the sequence of events that give rise to stars, planets, moons, asteroids, and comets. They also help us understand what we see when we happen to observe similar events going on in the galaxy today.

Given these generalizations gleaned from the study of the structure and composition of our Solar System, and reinforced by the apparently similar satellite systems of Jupiter, Saturn, and Uranus, we might sweepingly conclude that every star should have a similar system of planets. But the Sun is not representative of the average star in one important way: it is a single star. Most stars, perhaps as many as 80 percent of them, belong to multiple stellar systems. When formation conditions favor the origin of a double, triple, or even more complex multiple-star system, the fate and even the origin of planets may be seriously affected. This is an important matter that we shall return to in detail in Chapter 13.

It is certainly true that stars are not formed individually and in isolation: stars form from vast, massive interstellar clouds. Over a period of a few million years, a diffuse interstellar cloud of one thousand to many million times the mass of the Sun may be squeezed by light pressure from surrounding stars, or abruptly compressed by a passing shock wave from nearby exploding stars. The cloud of gases and dust may then become dense enough so that its own gravitational potential energy becomes comparable to the thermal energy of its component molecules. The cloud then begins a rapid contraction, essentially collapsing in free fall in less than a million years to a much smaller, denser cloud. As it collapses, the cloud repeatedly becomes unstable with respect to the growing gravitational force of its ever-denser component parts, and fragments into a number of smaller clouds, each of which, as it shrinks, soon fragments into a number of yet smaller and denser cloudlets, and so on until the individual cloudlets have masses comparable to a normal star. In this way, thousands to millions of stars all form together in a very small volume of space and in a very short interval of time.

When we look around us in our galactic neighborhood, at the spiral arm in which the Sun presently resides, we see several such star-forming regions of various ages and various stages of progress. The Orion Nebula is a present-day hotbed of star formation, with brilliant, highly active young stars seen still embedded in the dense gas and dust clouds from which they formed. The Scorpio-Centaurus cluster is older, the gas and dust largely swallowed by forming stars or blown away by the active young stars. The mass of this cluster has been so much reduced by gas expulsion that it can no longer hold itself together gravitationally. Its member stars are flowing out from its center like shrapnel from a bomb explosion, escaping from the "open" (gravitationally unbound) cluster of young stars and mixing into the population of older stars in our local spiral arm of the galaxy. Over the time of a single revolution of the galaxy (two hundred million years), as stars born together drift apart and migrate through other spiral arms, all visible connections between these stars will be lost and they will be thoroughly mixed in among the teeming billions of stars in the galaxy. Indeed, whenever we can identify a star-forming region, we can measure its expansion age, which is the time since these stars were all very close together in a compact, dense stellar nursery. (This is done simply by dividing the size of the cluster by the speed of the escaping stars.) That age ranges from a million to a few tens of millions of years for many different clusters.

The youngest clusters are enormously complex and dynamic. When we look at regions such as the Orion Nebula, where new stars are presently forming, we see an astonishing menagerie of strange phenomena. The young stars embedded in the nebula all have a distinctly higher luminosity than mature stars of the same color and temperature, with ultraviolet radiation making up a large percentage of the total. In many cases we can see direct evidence of an enormous, flat, warm, dusty gas disk in the equatorial plane of the star. Many also have clearly developed jets of enormously hot gas, traveling hundreds of kilometers per second, streaming out of both poles of the star like gushers in an oil field, spraying gas thousands of astronomical units into space. These hyperactive, hyperluminous stars are called T-Tauri stars, after the name of the first-discovered star of that type.

In older star-forming regions, in which the gases and dust have been largely blown away by the intensely active young stars and new-star formation is winding down, we see a cold, very thin dust disk in which bodies have grown to perhaps the size of asteroids. After ten million years or so, the young stars have settled down, lost their excess luminosity, ceased their excess UV light emission, and taken up residence as mature, sober *main sequence* stars. Planets in orbit around these main sequence (MS) stars have the best opportunity for long-term stability of illumination and heating, and hence for conditions conducive to the origin and development of atmospheres, oceans, and life.

For almost all of their lifetimes, all stars display a very particular relationship between their color (temperature) and their rate of energy emission (luminosity), ranging from very hot, blue, highly luminous stars all the way down to cool, red, extremely faint stars. Early in this century, Annie Cannon of Harvard College Observatory gave these stars designations that ran from O and B types (the bluest, hottest extreme) through A and F (white), G (yellow, like the Sun), K (orange), and M (red), based on color and spectral features. This continuous family of stars, first identified by astronomers Ejnar Hertzsprung and Henry Norris Russell in the early years of the twentieth century, has been known since that time as the main sequence. Originally, and mistakenly, astronomers believed that this sequence of stars, running down from highly luminous, hot blue-white stars (1 million Suns) to very faint, cool red stars (0.0001 Suns), was an "evolutionary" (age) sequence, in which the stars started out at the hot, bright end and slowly faded down toward the cool, faint end. But we soon discovered that the "evolutionary" interpretation was wrong: the main sequence is in sequence of mass, not age. The brightest, bluest stars may have fifty to one hundred times the mass of the Sun, but the faintest, reddest ones have only 7 to 10 percent of the mass of the Sun. The MS stars differ in mass: what they have in common is that they are all "burning" hydrogen fuel, fusing it by nuclear reactions in their interiors to make helium. Once the irresponsible youth of the star has ended, the star leaves the T-Tauri phase and settles down onto the main sequence. There it stays, moving only very slightly up the luminosity scale, for the entire time that the hydrogen fuel of the star lasts.

We know from laboratory studies of fusion reactions and from theory that a Sun-size star has an MS lifetime of about ten billion years. Stars on other parts of the main sequence have both different amounts of hydrogen fuel and different rates of fuel consumption, and hence may have very different life expectancies. A big, rare, O-class MS star with about sixty times the mass (amount of fuel) as the Sun gives off energy (burns its fuel) six hundred thousand times as fast as the Sun. It must therefore run out of fuel and leave the main sequence about ten thousand times as fast as the Sun, in a mere million years. A million years is about how long it takes for a young star to lose its nebula, turn its dust disk into planets, and settle onto the main sequence! Thus the brightest MS stars, of O and B classes, are very poor candidates for the origin of life: by the time they stabilize, they are dying.

Similarly, little, red, M-class MS stars, which are extremely common, may contain ten times less hydrogen fuel than the Sun, but are so miserly in burning it that they use it at a rate ten thousand times slower than the Sun does. They might therefore live as much as one thousand times as long as the Sun before running out of fuel (actually, they have trouble burning the little fuel they have, so they don't generally do much better than one hundred times the lifetime of the Sun). Our red cousins can therefore live as long as one trillion years, a million times as long as the blue-white spendthrifts. But these extremely faint, red M stars have their own problems: every MS star of classes G, K, and M has occasional brilliant stellar flares. Our Sun occasionally has a flare so bright that it increases the luminosity of the Sun by about 0.1 percent to 1 percent. Flares are even more important at lower masses: M stars can have their luminosity doubled or tripled by a bright flare. Planets in orbit around them will freeze most of the time and be severely scorched at rare and unpredictable intervals.

If planets are widespread, then each star that exhausts its supply of hydrogen fuel and leaves the main sequence may seal the doom of all its worlds. What actually happens to stars that use up their hydrogen fuel? The answer depends on their mass. Small stars, less than about two-thirds the mass of the Sun, simply fizzle out. Above two-thirds of a solar mass, the star is massive enough to contain extremely high temperatures and densities in its core; so high, in fact, that helium nuclei (the "ash" from hydrogen fusion in the star's core) can itself fuse together in threes to make carbon. A slightly larger star can continue the process a little beyond carbon, bringing four helium nuclei together to make oxygen. Progressively more massive stars can proceed onward by adding more helium nuclei to make neon, magnesium, silicon, sulfur, and argon. Finally, at around nine solar masses, the star can make atoms all the way up to iron and its neighbors. All these helium-fusing stars lie on their own "helium-burning main sequence." All former MS stars expand when they switch over to helium fusion, in the more extreme cases swallowing up their planetary systems as they expand. They are what we call giant stars. As direct results of their enormous radii, these giant stars have enormous luminosities and low surface gravities. They readily blow off much of their outer envelopes at speeds above escape velocity, shedding material that has been processed into heavier elements by nuclear reactions in their interiors. At and beyond a mass of nine Suns, the star can build up a core of iron and related elements, then detonate with a titanic explosion called a supernova. The exploding star may briefly, for a matter of weeks, outshine the entire galaxy in which it is located, with devastating consequences for worlds around nearby stars. The heavy-element debris from this explosion, liberally seasoned with radioactive materials synthesized in the blast, runs away at a few percent of the speed of light. The planets of such a star are immolated on the funeral pyre of their parent.

One of the ironies of the life histories of stars is that a considerable fraction of all the stars that become supernovas have masses so large that their main sequence lifetimes are less than the time it takes for a star-forming region to run its course. Therefore a large fraction of all supernova explosions occur in star-forming regions, where and while other new stars are being formed—*many stars die of old age while still in the nursery.* The blast waves from these titanic explosions sweep away the remaining unaccreted gas and dust in the cloud and cut short the star-formation process. But they also inject vast quantities of fresh radioactive elements into the disks surrounding many young nearby stars. These radioactive elements can then be accreted into growing planets, there to serve as a powerful heat source to melt and stir planetary interiors.

NOW THAT we know something of the nature and life histories of stars, we can look at the origin of stars with the origin of their planets in mind. It will be helpful to see how, at least according to present theories, the prestellar nebulae that spawn stars also give rise to dust disks, and dust disks give rise to planets.

The first question really must be, "Why do irregular, spheroidal interstellar cloud fragments shape themselves into flat disks?" The answer lies in the fact that these cloud fragments come together from different sources and directions at different speeds. It is exceedingly unlikely that the net result of this random growth-and-collapse process will have no net rotational motion. But the pieces of the cloudlet, because of their different histories, have not yet shared their energy and momentum with each other: in effect, different parts of the cloud move independently, being as yet unaware of what the net rotation of the cloud will be. As the cloud shrinks, collisions between parts of the cloud become more frequent, and eventually the parts all "agree" on where the rotation axis is, which direction the cloud is rotating in, and how fast it is rotating. Very slowly rotating clouds may collapse almost unimpeded, leading quickly to very massive stars that age extremely rapidly. These are pre-supernova stars. The most rapidly rotating cloudlets tend to split into two or more parts, each orbiting around their common center of mass. These separate parts may then evolve independently, giving rise to a gravitationally bound cluster of two, three, four, or more stars. In the wide middle ground of moderate rotation speeds, medium-size single stars — or close double-star systems — can be made.

Each of these cloudlets, once isolated, continues to radiate heat. Loss of heat from the interior of the cloud lowers its internal pressure (which supports the outer part of the cloud), and the cloud collapses further. This generates more heat, which must in turn be lost to permit further collapse. But each cloudlet is rotating around a well-defined axis. Collapse toward the spin axis is hindered by the rotational motion of the cloud: any collapse that occurs causes the cloud to spin faster, which further opposes collapse. On the other hand, collapse parallel to the spin axis, toward the cloud's equator, has no such difficulty. As a result the cloud quickly collapses, as observed, into a flattened disk in the equatorial plane of the system. The densest central part of the disk collapses eventually into a star (or a close binary pair). The dust in the disk settles down onto the equatorial plane, forming a very thin layer where its concentration is so high that the dust readily clumps together. This agglomeration is the cosmic equivalent of what happens under your bed when you're not looking: airborne dust settles to the floor, where slow air currents stir the dust. Dust particles, upon gentle collision, stick together to make fluffy domestic (or cosmic) dust-bunnies.

The cosmic dust-bunnies continue to collide until asteroid-size bodies, compacted by countless collisions, form a sort of superdense, systemwide asteroid belt. Only then does gravitation begin to play an important role in their growth.

Solid bodies embedded in a gas disk will naturally travel around the Sun at a different speed than the gas disk itself: the disk is partly supported by the pressure gradient in it, whereas the solid bodies are far more strongly affected by gravitation than by gas pressure forces. The difference between the speed of the gas (slow) and the large solid bodies (progressively faster, depending on their size) causes the big bodies to suffer frictional loss of energy and slowly spiral inward. Big and small solid bodies, experiencing different proportions of these forces, evolve inward at different speeds and therefore tend to collide with each other at low speeds. The smaller bodies get swept up by the big ones. Continuing accretion of these asteroidal bodies builds up over time a new population of bodies up to about the size of the Moon.

Far from the Sun, in the cold, dark outer nebula, the temperature is so low that particles of ices are present along with the rocky mineral grains. But, because the gas is so cold (near 100 K; −275°F), its molecules are traveling much slower than the same molecules would be closer to the Sun, where the nebula is much warmer, near 1000 K (1340°F). Solid bodies far from the Sun have more mass available to them because the ices are condensed there. They will grow to lunar, or Ganymedian, size in that cold gas. But at that point, quite abruptly, the gravity of a growing solid "dirty snowball" reaches the point at which it can gravitationally capture cold hydrogen and helium gas from the nebula in which it is embedded. It then begins to grow much more rapidly, and (because of the large surrounding envelopes of captured gas) begins to collide much more frequently with others. The largest of these grow to make planet-size bodies with captured gas envelopes. The smallest get captured by the largest.

In the Jupiter-Saturn-Uranus-Neptune region, the highest concentrations of solids (and therefore the most rapid collision and accretion rates) occur closest to the Sun. Even closer to the Sun, ice is evaporated, the solid bodies have about 50 percent less mass, and the nebula is warmer and therefore harder to capture. Thus there is a fairly well-defined point in the nebula at which the formation of giant, gas-rich planets becomes possible. Jupiter forms near that point, separating rather quickly from the nebula and rather efficiently capturing the available hydrogen and helium. The resulting planet has a composition close to that of the Sun, but still a little short on hydrogen and helium because the capture of these gases was not perfectly complete. Saturn, a little farther out, does the same thing, but more slowly: less mass of solids is available, and the longer orbital periods farther from the Sun make collisions less frequent. Growth and gas capture are even less efficient at the distance of Uranus and Neptune.

At some point in time the Sun ignites in the center of the system, and the nebular gases that have not fallen into the Sun or been captured by a giant planet are stripped away by the Sun's T-Tauri phase. The giant planets quickly have their supply of gas cut off. The giant planet closest to the Sun (Jupiter) has by this time captured the most, with Saturn next, and Uranus and Neptune lagging far behind. At the time of the dissipation of the nebula, the four giant planets were all largely assembled. But much of the mass of Uranus and Neptune must have still resided in dirty iceballs that were too small to hold hydrogen and helium when the nebula went away. When these iceballs are swept up by a planet, they bring in abundant ice and rock, but no hydrogen and helium. The final sweep-up of these solid bodies must take a very long time; indeed, the Kuiper belt is probably made up of bodies from the Uranus-Neptune region that have still not been swept up. Interestingly, all four giant planets have similar core masses (of icy and rocky materials), even though their hydrogen and helium contents drop off precipitously as we go outward from the Sun.

Meanwhile, the sweeping of residual gas and dust from the inner Solar System must have been even more efficient than it was beyond Jupiter. Gas drag would cease to be a factor as the T-Tauri Sun ejected it from the system. The interactions of the Moon- and asteroidal-size bodies in the terrestrial planet region must thereafter have been exclusively gravitational. If this were the whole story, collisions would thereafter be exceedingly rare and the growth of the planets would be arrested at an embryonic stage. But the Russian physicist V. S. Safronov pointed out nearly fifty years ago that, in such a swarm of solid bodies, gravitational "stirring" by their mutual interactions will result in all the bodies approaching a limiting speed that is set by the escape velocity of the largest body in the swarm. This stirring provides enough random velocity to pump up the inclinations and eccentricities of their orbits, so that the bodies in the swarm must very frequently cross each other's orbits. Collision and accretion again become possible. Growth continues up to the size of the present planets. Each accretion event averages out the motions of the colliding bodies; each near miss tends to build up the eccentricities, inclinations, and relative velocities of the bodies involved. If the velocities get too high, growth is again interrupted: the collisions become so violent that they are disruptive.

Once a planet is formed, various forms of internal activity may become important. Indeed, the very accretion process brings in large amounts of heat. It is likely that the growing planets began to melt in their interiors before they grew to the size of Mercury. Other early heat sources such as decay of short-lived radioactive isotopes (especially aluminum-26), despinning of a planet (such as Mercury) by tidal forces, or electrical-induction heating by the T-Tauri-phase dense solar wind, may also favor early melting. Melting allows dense minerals and melts, especially metallic iron-nickel and iron sulfide, to sink to the center of the planet to form a dense, electrically conducting core. Dense silicates rich in magnesium and iron oxides, because of their high melting points, are either incompletely melted or are the first to solidify during the cool-down phase after accretion has passed its peak. These dense minerals sink to the bottom of the melted rocky region to form a mantle. The residual melt, mainly composed of "incompatible" elements that do not fit readily into the crystal structure of the dense iron-magnesium silicates, has a low melting point and a low density and therefore is extruded upward, where it eventually erupts onto the surface and cools to form the crust. Highly volatile materials, such as water brought in by hydrated salts and water-bearing minerals, or carbon dioxide made by heating carbon and iron oxides together, are released from the interior of the planet as volcanic gases.

Continuing geological activity is dependent upon a continuing heat source, one that energizes a planet for billions of years. The most important such heat source is the radioactive decay of isotopes of potassium, uranium, and thorium, which are among the distinctive products made in supernova explosions. In special circumstances, such as on Jupiter's moons Io and Europa, tidal interactions can also sometimes keep the internal pot boiling for billions of years. In that system, gravitational interactions of the four large Galilean satellites keep pumping up the eccentricities of their orbits. The eccentricity of each satellite's orbit introduces a mismatch between its (constant) rotational speed and its (variable) orbital speed. Tidal interactions with Jupiter then damp out the eccentricity and convert the extra energy into heat, which melts solids and permits dynamic change. *The conditions of planetary genesis often guarantee that thermal evolution takes place.*

Whatever the sources of heat, thermal evolution in turn melts minerals, drives eruptions that deliver water and gases to the surface, extrudes low-density volcanic magmas that solidify into crustal rocks, and lets the densest components of the body, such as metals and sulfides, sink to the center to form a core. Geochemical evolution of the crust and the inorganic chemical evolution of the atmosphere occur in concert on planetary bodies that grow large enough to melt. The products of inorganic chemical evolution include vast numbers of relatively simple organic molecules, among them the building blocks of life.

There are many clues to life's role in the planetary drama, and many possible routes to its origin. Organic matter, which is essential to life as we know it, is incredibly widespread in nature. Giant interstellar clouds, which spawn stars and planets, are rich in organic molecules. Radio astronomers have found dozens of organic compounds, some with over a dozen atoms, in these so-called giant molecular clouds. Many of these molecules are highly reactive: at higher temperatures and pressures, and especially in the presence of liquid water, they spontaneously combine to make even more complex molecules and polymers.

Interestingly, many of the same molecules have been found in comets. Astronomers examining the spectra of comet heads and tails have found a host of small organic molecules that strongly suggest that cometary material is virtually unchanged presolar material derived from interstellar dust. Among them are the important building blocks acetylene (C_2H_2), hydrogen cyanide (HCN), and formaldehyde (CH_2O).

In 1986, at the time of the much-heralded return of Halley's Comet, the spacefaring nations had a unique opportunity to send probes to study a bright and active comet that had been under study for centuries. The

Soviet Union sent two heavy *Vega* (for Venera-Galley, meaning Venus-Halley in Russian) spacecraft to swing by Venus and use Venus's gravity to divert them toward Halley. The European Space Agency sent its first planetary mission, the *Giotto* spacecraft, to fly by Halley's nucleus and examine it at close range. Japan likewise undertook its first planetary mission, an automated probe named *Sakigake* (all Japanese spacecraft are named after flowers) to study the comet's atmosphere. The only nation with the technical ability to carry out such a mission that declined to do so was the United States.

One of the most interesting discoveries during this time was made by the mass spectrometer experiment conducted by the *Giotto* mission. It found that the gases evaporated from Halley's icy nucleus contained a long-chain molecule, a polymer of the simple organic molecule formaldehyde, CH_2O, with the structure $-CH_2-O-CH_2-O-CH_2-O-$, etc.

Comets originate from the outer Solar System, where temperatures during the nebular disk phase were not high enough to destroy the organic matter inherited from interstellar clouds. Some of the ices in comets may have been evaporated and recondensed; other comets may be made of solid materials that have remained unchanged since before the Solar Nebula itself was formed. Indeed, many comets may be bodies ejected along with the Kuiper belt "iceteroids" from the region in which Uranus and Neptune swept up local solids, long after the demise of the nebular disk.

Closer to the Sun, and under the influence of disturbing chemical processes such as lightning, shock waves, and ultraviolet light from the T-Tauri phase of the Sun, organic matter may have been synthesized in large quantities within the nebula. The outer half of the asteroid belt today bears the signature of abundant tarry organic polymers. The same materials make up as much as 6 percent of the mass of the black, volatile-rich meteorite class called the carbonaceous chondrites. These meteorites contain many volatile organic compounds, some inherited from the interstellar cloud out of which the Solar System formed, some made in the Solar Nebula, and some within the asteroids or extinct comets within which the meteorites formed. Radioactive decay dating and textural studies show clearly that the carbonaceous meteorites have not been significantly heated since 4.55 billion years ago, when the Solar Nebula was present and the planets had not yet formed.

But organic matter frozen up in comet nuclei or distant asteroids is of little biological interest. Can this organic matter be made available in locations, such as the surface of Earth, where life can actually arise? Does the ubiquitous presence of organic matter in interstellar space guarantee that every young planet will be seeded, via comet impacts, with the building blocks of life?

The most obvious way to introduce organic matter onto a sterile young planet is to send a comet full of organic molecules crashing into it. A dozen or so kilometer-size comets pass through the inner Solar System every year, each with a probability of one in a billion of striking a terrestrial planet. One comet this size strikes Earth about once in every hundred million years. But this approach to importing organic matter is a woeful failure: the comet is so fragile, and the impact so violent, that the result of the impact is a fiercely hot (about one hundred thousand degrees) explosion fireball that completely destroys every molecule in it. But all is not lost: as pointed out by Chris Chyba of the University of Arizona, the dust shed by the comet is a different matter entirely. Tiny particles of fluffy dust released by the evaporation of cometary ices create a huge dust cloud surrounding the icy nucleus and stretching along its orbit. Whether the intact nucleus strikes a planet or not, the dust from its tail very likely will. And these fluffy little particles act like tiny parachutes, slowing down gently at high altitudes without ever getting hot enough to vaporize or burn up their precious organic cargo. Larger pieces of cometary dust, which burn up in the upper atmosphere, are called meteors. A meteor shower is nothing more than Earth passing through the dust stream emitted by a comet and sweeping up many particles in a short time. An excellent example is the Leonid meteor shower, a dense dust cloud shed long ago by Comet Tempel, which sweeps Earth for about one hour every thirty-three years. When it hits, the rate of fall of meteors visible to the unaided eye reaches about 150,000 per hour (40 per second!). The next Leonid meteor storm is due in November 1999.

Another source of organic matter on planets is the fall of carbonaceous chondrite meteorites. Studies of fireballs have made it clear in recent years that the large majority of the carbonaceous meteorites that strike Earth's atmosphere is utterly destroyed. But a small percentage not only make it through to the ground intact but are found before they can be destroyed by rain, and delivered safely to museums. Of the entering bodies that successfully drop meteorites on the ground, only about 1 percent of the original mass may survive passage through the atmosphere.

Planets can also produce their own organic matter, provided only that the atmosphere is devoid of free oxygen gas. Oxygen readily attacks and destroys a wide variety of organic molecules. But oxygen is not an early component of planetary atmospheres: oxygen is abundant only on Earth. Here its presence is due to the effects of plant photosynthesis, a process that was clearly irrelevant before the origin of life. Before oxygen, such widespread inorganic processes as lightning and thunder, radioactivity, shock waves from entering meteors, and exposure to solar ultraviolet radiation can rearrange the atoms in simple molecules and produce more complex ones. Lightning passing through a mixture of nitrogen, water vapor, and carbon monoxide can make simple water-soluble organic compounds. Even better results obtain when methane and ammonia are present at the start. The product compounds, washed into the rivers, lakes, and oceans by rainfall, are removed from exposure to lightning and ultraviolet light, and can then react with each other in relative safety to make more complex molecules. Consider only the molecules acetylene, formaldehyde, hydrogen cyanide, and water: reactions between these simple, ubiquitous compounds can make sugars, hydrocarbons, organic acids, organic bases, and amino acids. But to do so, liquid water is essential.

Organic matter, whatever its origin, can do marvelous things in water. The simple building blocks of sugars, amino acids, and organic bases, once made by reactions in liquid water, are then available for incorporation into much larger molecules, much in the way that a one-room brick factory makes twenty-story brick buildings possible.

All this organic chemistry does not occur in splendid isolation: on Earth, for example, it occurred on the surface of a chemically complex rocky planet. On the surface of such a planet, ferrous iron minerals, clay minerals, sulfides, phosphates, and a host of other inorganic species may be present. Ferrous iron (an iron atom with two electrons removed, making an ion with a double positive charge, Fe^{++}), for example, which is almost ubiquitous in rocky Solar System bodies, has the interesting property that it can protect organic matter from oxidation by low concentrations of oxygen. Many clay minerals can not only attract and adsorb (bind to their surfaces) many simple organic molecules, but in some cases they can also help these simple molecules polymerize into complex products of biological interest. Phosphate minerals dissolve to some extent in water and can bind to sugars. Sugars can bond directly together to make long chains called polysaccharides, of which starch is a familiar example. Sugars can also bind, in concentrated solutions, to organic bases. When all three of these components are present, units called nucleosides (each consisting of one organic base molecule bonded to one sugar molecule) can be linked together by phosphates into extremely long polymer chains called nucleotides. One such nucleotide is DNA, the molecule that carries genetic information. Another is RNA. Organic acids with long hydrocarbon chains in them (fatty acids) cooperate to make lipid bilayers, which are familiar to us as cell walls. Amino acids can link together to make short chains called peptides, or long chains called proteins. Many of these peptides and proteins, often those containing metal ions weathered out of rocks, serve as efficient catalysts that greatly accelerate the rates of important biochemical reactions. These organic catalysts are called enzymes.

Self-replication begins on the inorganic level. The surfaces of some types of clays can not only serve as a template for the assembly of complex organic polymers, they can also release the product and make another in the same template. Even more astonishing, if sheets of atoms are cleaved off the surface of such a clay, the separated clay sheets may capture ions out of solution to grow new layers that are exact copies of themselves. Even flaws in the pattern can be replicated, demonstrating the ability of clays to "mutate" as well as replicate. It may well be that organic matter was first replicated by and on inorganic clays, and that organic matter learned how to replicate from these clays. From the primordial red clay, life was first shaped. The Hebrew word for red clay is *adamah*.

But this magic does not happen everywhere. If the temperature is too low, the solutions freeze solid, the dissolved molecules are immobilized, and reactions cease. If the temperature is too high, the delicate organic molecules are destroyed. In short, if there is no liquid water, nothing of interest happens. If there is oxygen present, the organic matter oxidizes to carbon dioxide, nitrogen, and water. And if there is no source of high-quality energy, none of the building blocks are made.

Life is not for every planet—at least, life as we know it is not suited to exist everywhere. Planets that follow very eccentric orbits may alternatingly freeze and roast their passengers; planets that don't rotate may have almost no stable life zone between the dayside furnace and the nightside deep freeze; planets that closely approach other planets may develop chaotic orbits that cause wild temperature excursions or even collisions; planets orbiting multiple stars may not survive long enough to develop life. It's a tough universe out there.

We should also not forget that planetary surfaces are to a great measure dependent on astrophysical conditions: a star that suddenly enters the red giant phase and swallows its planets, or one that quickly goes supernova and vaporizes its planets, is obviously unsuited to planetary life. A star that frequently fires off brilliant flares, shot full of killing ultraviolet radiation, may not be a good bet. A star that, at random intervals, triples its luminosity and scorches its planets, is probably even worse.

Clearly, we need to explore the varieties of known worlds and suns, understand what makes them more or less desirable as abodes for life, and then apply that knowledge to other planetary systems. Perhaps then we can identify user-friendly stellar systems from afar.

4

RECIPES FOR A ROCKY PLANET

THE ONE WORLD WITH WHICH we are all personally familiar, Earth, is but one of a variety of rocky worlds accessible to our telescopes and spacecraft. Our Solar System actually offers no fewer than six credible rocky worlds for inspection: Mercury, Venus, Earth, our Moon, Mars, and Jupiter's innermost major satellite, Io. Each planet contains ninety-odd chemical elements arranged into a distinctive suite of minerals and a characteristic set of rock types, layered into a unique structure. The number of known minerals is about three thousand, of which all but about one hundred are known only on Earth. But fortunately, as diverse and complex as these bodies are, we can understand an enormous amount about them by looking at only six simple, familiar categories of materials. Each of these materials existed in the Solar Nebula out of which the Sun and its retinue formed, but each of them also can exist in contact with a solar-composition gas only within a specific range of temperatures: get them too warm, and they evaporate; too cold, and they react with, and are destroyed by, other gases such as water vapor. Some of these materials, called *refractories*, have very low vapor pressures and are easily condensed. Most of these refractory materials are also hard to melt.

Very close to the center of the nebular disk, where temperatures are very high, the solar-composition mix of elements will be completely vaporized and mixed into a uniform gas. Like the Sun, this gas is mostly hydrogen and helium, with about 1 percent water vapor, a fraction of a percent of carbon monoxide and nitrogen, and about half a percent of potential rock-forming materials. The latter are vapors containing silicon, magnesium, iron, calcium, aluminum, sulfur, nickel, potassium, and about eighty other minor and trace elements, which are kept from condensing by the very high temperatures near the center of the nebular disk.

But a little farther out from the center, at slightly lower temperatures, the most refractory minerals will condense. The mix of gas and solids found there still has the same overall composition, that of the Sun itself, but that material is now divided between gas and condensed refractory minerals. The solids can stick together and accrete into larger solid bodies, beginning the chain of accretion events that leads eventually to the formation of full-size planets. The refractories generally have high contents of calcium, aluminum, titanium, and a host of rarer elements that are chemically similar to these, including the important radioactive elements uranium and thorium. None of these refractory elements are nearly as abundant as silicon, magnesium, and iron: the refractories collectively amount to only a small fraction (about 5 percent) of all the possible rock-forming material. But that 5 percent contains about 50 percent of all the available radioactive heat sources.

Another step outward from the center of the nebula takes us to a region just cool enough for iron metal to condense. Iron is very abundant, making up about 30 percent of the mass of most rocky planets (and as high as 60 percent for Mercury). Certain other elements that are chemically similar to iron, especially nickel and cobalt, condense along with iron to form a natural stainless steel alloy closely similar to that found in iron meteorites. Numerous rare elements with a chemical affinity for iron also enter into this metal, notably the platinum-group precious metals and a group of nonmetals, including gallium, germanium, arsenic, and small fractions of phosphorus and carbon. The solid material in the region where metal is freshly condensed then consists of a mixture of the only two classes of material that are condensed there, refractories and metal, in the same proportions in which they are present in the Sun (0.05:0.30). Thus the condensed solids are about 15 percent refractories and 85 percent metal. The accompanying gas consists of everything else, all the materials that remain uncondensed at the local temperature.

At temperatures only slightly lower than the condensation point of metallic iron alloy, magnesium silicates also condense. Because magnesium and silicon are abundant, the magnesium silicates make up the majority of all possible rock-forming material. The magnesium silicates are perhaps the least familiar materials on our short list. That is because they are only minor ingredients of Earth's familiar crust, and are instead concentrated in the mantle, where we can't see them. The main example of this class is the mineral enstatite, an entirely forgettable colorless crystal.

Over a temperature range of several hundred degrees, the only materials condensed are the refractories, the iron-nickel metal, and the enstatite. As simple as this list appears, the large majority of the rock-forming elements are already contained in these solids. Almost all of these elements, except silicon, magnesium, and iron, are so rare in nature that they make no difference to our story—except for the few that are important because they are radioactive. At this point, the large majority of the mass of potential rock-forming material is condensed. Planets that accrete from this mix of solids will be full-fledged bodies lacking only low-temperature components, those with high volatility and low condensation temperatures. The proportions of refractories, metal, and silicates are now roughly 0.05:0.30:0.65.

Continuing outward from the Sun, we encounter ever-cooler gas. After several hundred degrees of further cooling, a new class of elements begins to condense. Over a further range of a few hundred more degrees, these relatively volatile rock-forming elements condense into minerals. The three most important of them are sodium, potassium, and sulfur. Several minor elements that are chemically similar to these also condense, but they are so rare as to be of very little interest to us. During cooling, sodium and potassium vapor react with refractory calcium and aluminum compounds to make one of the most familiar families of minerals known to residents of Earth, the feldspars. Almost any random rock picked up on Earth's surface contains one or two feldspar minerals. Next comes the element sulfur, which, at low temperatures, reacts with metallic iron to make a sulfide called troilite. Troilite is rare on Earth's surface because Earth's oxygen and water react with it readily, weathering it into iron oxides and sulfates. But troilite is nearly ubiquitous in meteorites, which come from bodies (asteroids, the Moon, Mars, and so on) that do not have moist, oxygen-rich atmospheres.

Note that the formation of both of these groups of minerals requires that the cooling nebular gas be in contact with minerals that formed at much higher temperatures: feldspar formation requires reaction of calcium-bearing refractories with cool gas containing sodium and potassium vapor, and troilite formation requires contact between metallic iron and cool gas containing sulfur compounds. If for any reason the early condensates and the gas are not in contact with each other (for example, if the early-condensing materials are buried by accretion of later condensates), then these reactions cannot occur and no feldspars or sulfides are formed. Sodium, potassium, sulfur, and their rarer cohorts would then be left stranded in the cooling gas until they can discover some other way to condense.

Iron sulfide formation cannot use up all the available iron metal, because sulfur is much less abundant than iron. Iron, however, can not only react with sulfur-bearing gases (mainly hydrogen sulfide), but also with water vapor present in the nebula. As the temperature falls, iron oxides become more and more stable. The product of oxidation of iron metal (Fe, from the Latin word for iron, ferrum) by water vapor is ferrous oxide, simply FeO. But, by an interesting coincidence of nature, FeO can substitute quite freely for magnesium oxide, MgO, in complex minerals: the sizes and charges of the Mg^{++} and Fe^{++} ions are virtually indistinguishable. Therefore whenever FeO starts to form, it immediately enters into a variety of magnesium-bearing minerals in place of magnesium oxide. Iron oxide then becomes a minor component of enstatite. As the temperature falls (that is, as we move farther outward from the Sun), enstatite, which started out as virtually pure $MgSiO_3$, takes on ever more FeO. Pure enstatite is the iron-free extreme of a mineral family called the pyroxenes. Ferrous oxide, however, has a clear green color, a very distinctive tint that was for many decades the distinguishing feature of Coca-Cola bottles.

Oxidation of iron has several other consequences. First, the metal that is being oxidized is not pure iron: it is an alloy containing nickel, cobalt, platinum metals, and so on, all of which are harder to oxidize than iron. Removal of iron raises the concentrations of all the other ingredients of the metal grains. The nickel concentration in the remaining metal, originally only 7 percent at high temperatures, can reach 60 percent at low temperatures when most of the iron has been oxidized. Finally, adding major amounts of FeO to enstatite not only makes the mineral get progressively greener, but also changes it into a new mineral. The reason is simple: iron, magnesium, and silicon have very nearly equal abundances in solar material. Suppose we consider one hundred atoms of each. At the temperature at which enstatite condenses, we have one hundred molecules of $MgSiO_3$ (one MgO per silicon) and one hundred atoms of Fe metal. When the temperature drops low enough to make ten molecules of FeO (leaving ninety atoms of Fe metal), the FeO enters into enstatite *in place of* some of the MgO: there is now more than one metal oxide molecule per silicon. Some of the enstatite picks up FeO to make an FeO-bearing pyroxene, but some is also altered into a new mineral called olivine, which has two metal oxide molecules per silicon, $(Mg,Fe)_2 SiO_4$. Olivine with enough FeO in it to produce an attractive green color is the semiprecious gem peridot. At low temperatures, with iron metal almost completely converted to FeO (and FeS), almost all the pyroxene has been made into olivine.

These FeO-bearing minerals are important during the early evolution of organic matter, because FeO will preferentially react with any oxygen that comes by, thus protecting the organic matter from oxidation. Ferrous oxide is oxidized further by oxygen to make the black oxide magnetite (Fe_3O_4), and then to the rust-red oxide hematite (Fe_2O_3). Once all the iron has been oxidized to hematite, it loses its ability to absorb any more oxygen, hence its ability to protect organic matter from oxidation. Even worse, when hematite is heated with a small amount of organic matter, it gives up some of its oxygen and "burns" the organic matter. Thus the degree of oxidation of iron is of great biological significance.

One critical component of a biologically interesting planet still remains missing from our inventory of preplanetary solids. That is water. Providing a young planet with ample water and other volatiles is an essential prerequisite for the origin of life as we know it. At rather low temperatures, while there is still some metal present and FeO is rather abundant, water vapor in the nebula can begin to react with dry silicates to make water-bearing minerals that contain hydroxyl, OH. Two simple examples of such a mineral are brucite, $Mg(OH)_2$, and goethite, FeOOH. When brucite is heated, it gives off water vapor, leaving solid MgO. Goethite likewise gives off water and makes hematite. Water may also enter minerals as water of hydration, such as in $CaSO_4{\cdot}2H_2O$, the dihydrate of calcium sulfate, which bears the mineral name gypsum. All such minerals dehydrate at high temperatures. Therefore only minerals formed at low temperatures can bring in water to be incorporated into growing planets. These water-rich, low-temperature minerals include a number of clays. The most highly oxidized meteorites, those that contain magnetite and no metallic iron, also contain abundant bound water in minerals such as serpentine. In the most extreme cases, as much as 20 percent of the solid body may be chemically bound water (hydroxides or hydrated minerals). In rocky planets, it is these low-temperature minerals containing bound water that are the source of most or all of the water vapor and liquid water that will end up on the surface of the planet when it melts and releases its volatiles in volcanic eruptions.

In addition, cooling to far lower temperatures can freeze out large quantities of water vapor in the form of ordinary ice. Far enough from the Sun, where temperatures in the nebula are low enough for ice to condense and snow out, water ice can be the most abundant single mineral. It is in such regions that ice-plus-rock bodies, such as comet nuclei and ice-bearing satellites, must form. Impacts of such ice-rich bodies can also be an important source of water for the rocky planets.

This "condensation sequence" of minerals (the order in which minerals become stable as one moves outward from the center of the Solar Nebula disk) strongly influences what minerals are available at each distance from the Sun. Each small preplanetary solid body, as it grows, efficiently captures nearby solids, but less efficiently captures material that originated far from its orbit. Only in the final stages of planetary accretion, when lunar-size bodies are perturbed into more eccentric orbits by the gravitational effects of rapidly growing planets, may they stray widely in the Solar System and collide with a distant planet. Thus even though this mineral sequence may be an excellent guide to what raw materials were originally present at each distance, random processes can nonetheless significantly affect the compositions of the planets that ultimately accrete from these raw materials.

The possibility of long-range migration of early nebular solids is greatly enhanced by the presence of the giant planets, especially Jupiter. So powerful is its gravity that it can stir up bodies in resonant orbits (orbits whose periods are harmonically related to Jupiter's) throughout the asteroid belt. Jupiter's perturbing force is the major cause of asteroids crossing Earth's orbit, and the controlling factor in the orbits of most of the short-period comets. Both these ejected belt asteroids and short-period comets are potent suppliers of material that originated far out in the nebula, at low temperatures. Comet nuclei, for example, formed at such low temperatures that a variety of ice-forming materials (water, methane, ammonia, carbon monoxide, carbon dioxide, and so on) are condensed along with the rock-forming solids, making the typical comet nucleus a veritable "dirty snowball."

It may be that these rare impactors from distant, cooler climes are actually the main source of volatile, easily vaporized components (hydrogen, carbon, nitrogen, and noble gases) in Earth and the other terrestrial planets. Alternatively, these volatiles may have been present in the original preplanetary solids near Earth's orbit, as carbon and nitrogen dissolved in iron-nickel metal in small but adequate traces. Earth and Mars provide no convincing evidence of a cometary origin of their volatiles, but Venus, with a vastly larger endowment of noble gases than either Earth or Mars, may point to just such a source. Perhaps a truly gigantic cometary body, such as a straying Kuiper belt iceball kicked inward by Uranus or Neptune, provided these gases in a single random event.

The present infall rate of comets and asteroids on Earth and Mars (where the explosion of an impactor can blast itself back out of Mars's gravity well) would not contribute a significant fraction of the observed volatiles even over 4.5 billion years. But the impact rate of cometary material may have been much higher during the era when Uranus and Neptune were struggling to clean up the small solid bodies near their orbits. Interestingly, that same present infall rate does appear sufficient to provide most of the other atmospheric gases (except carbon dioxide) on Venus.

WE PICTURE each planet, then, as being dominated by material of local origin but seasoned by significant amounts of solids that originated much farther from the Sun. Virtually all the mass of the terrestrial planets was accreted by violent impact events in a relatively short period of time, perhaps one hundred million years. All this material would generally be very well mixed by the impacts, leading to fairly homogeneous growing planets.

At some point, possibly before each planet reached 1 percent of the mass of Earth, the heat given off by infall of large solid bodies began to melt the young planet. Once melting begins, minerals and melts readily differentiate into layers of different composition according to their mutual solubility and density. The behavior of the chemical components of the planets during differentiation is fairly well understood, thanks to studies of rocks from Earth, the Moon, Mars, and meteorites from a number of differentiated asteroids, and to a wide range of laboratory experiments. The basic scheme of how the elements sort themselves out is rather simple. First, when a planet melts, it readily forms a very dense, mobile liquid consisting of liquid iron and liquid sulfides dissolved in each other. A mixture of iron and sulfur, like a mixture of salt and ice, begins to melt at a temperature far below the melting point of either pure component. This phenomenon is called *eutectic melting*. (Other minor ingredients of the metal-rich melt, such as nickel, cobalt, phosphorus, carbon, and oxygen, tend to lower the eutectic temperature even further.) The iron-iron sulfide liquid, with its very high density, naturally sinks to the center of the planet, forming a core.

As the core materials fall in the planet's gravitational field, they liberate a considerable amount of gravitational potential energy, which appears as heat to melt more material. Later cooling may partially crystallize this liquid, leading to a separation of iron-nickel metal (the densest) and a liquid rich in iron and sulfur. These two layers correspond roughly to Earth's solid inner core and liquid outer core. Many chemical elements that are found as free metals or sulfides are dissolved in the core-forming melt and extracted very efficiently from the rest of the planet. The elements that enter readily into sulfides are called *chalcophiles* ("sulfur-lovers") and those that follow the native metals are called *siderophiles* ("metal-lovers"). Note that a planet made of relatively high-temperature materials, nearly devoid of iron sulfide (Venus?), will have to be heated not just to the Fe-FeS eutectic point to begin to melt, but all the way to the melting temperature of iron, hundreds of degrees higher. A liquid iron core, once formed, may freeze to leave virtually no residual sulfide-rich melt, and hence no liquid outer core within which currents can generate a planetary magnetic field. Likewise, a planet formed at even lower temperatures (Mars?) may have virtually all of its iron oxidized to FeS and FeO, and have a core made of sulfides.

By the time the metal and sulfides have been buried in the core, the rest of the planet has become extensively melted. This outer portion, composed of molten silicates, is said to be made of *lithophile* ("rock-loving") elements. Recall that, because of the abundances of the elements, the large majority of the rocky material of the planet consists of silicates of magnesium and iron. Olivine has a very high melting temperature and is the densest major component left after core separation. Therefore as the silicate portion of the planet begins to cool, olivine begins to crystallize first, and sinks to rest on the surface of the core. The next to crystallize is pyroxene, which is somewhat less dense than olivine but still denser than the melt. Further cooling leads to the formation of a mantle of olivine and pyroxene, containing most of the mass of silicate materials, which have sunk in the ocean of molten rock. The residual melt contains those elements that do not fit well into the crystal lattices of olivine and pyroxene, either because they have the wrong ionic charge (such as sodium, potassium, and aluminum) or the wrong ionic radius (such as barium or uranium) to substitute for iron, magnesium, or silicon. These *incompatible elements*, along with the volatile *atmophile* ("air-loving") elements, rise to the surface and generate the crust, oceans, and atmosphere. The crust is characterized by very high concentrations of silicon dioxide, aluminum, calcium, sodium, and potassium, and is strongly enriched in the whole range of other incompatible rock-forming elements. Its dominant minerals are quartz, feldspars, and the like. Attack on these minerals by liquid water and air extracts soluble materials from the rocks, flushes salts into the oceans, and leaves behind a fine-grained residue rich in clays.

How well a planet does all these things depends not only on its composition but also on its size. Small planets cool very readily and have a difficult time sustaining geological activity. They easily lose light gases from their atmospheres. Large planets, because they have less surface area relative to their mass, keep warm more effectively, and collect thicker layers of atmosphere and oceans. They also have higher accelerations of gravity. High gravity limits the height to which mountains can be raised, and flattens the planetary surface. Therefore, planets of identical composition and location may end up looking very different if they have different masses. We must look more closely at the consequences of the formation of planets of very different size.

MORE OF THE SAME IS NOT THE SAME

IN STUDYING PLANET FORMATION, WE first asked where the material to build a planet comes from and what it's made of. Then we asked how the raw material of a planet is put together. Now we need to know how the planet's size is determined.

The beginning cook learns that, if the proportions of ingredients in a recipe are kept the same, the recipe can readily be scaled up or down. The professional cook, however, knows that this is not strictly true: roasting a three-thousand-pound apatosaurus haunch would be by no means the same thing as roasting the leg of a Cornish hen, and baking a one-thousand-pound cake is very definitely not like baking a cupcake. Size matters.

Imagine a set of a dozen planets that at their inception all have exactly the same composition and orbit, but different sizes. Would they all turn out pretty much the same? Not a chance. Just as in the National Football League, size is important. The smart, agile 300-pound lineman will, we predict with confidence, have a longer active career than the 150-pound lineman, no matter how smart and agile. Of course, a player's position is also important: a 150-pound placekicker who can boom in field goals from the next town over may cut out quite a niche for himself (assuming that he doesn't have too many collisions with 300-pounders). A lot depends on what you're made of, a lot depends on how big you are, and a lot depends on your position. And so it is with planets.

How does size matter? It first comes into play in the youth of the Solar System, when the planets are still growing and there is still a lot of mass wandering around in small, unaccreted bodies. Large bodies, as we have seen, stir up small bodies by means of their gravitational acceleration as they pass by, giving the small bodies more eccentric orbits. But small bodies in eccentric orbits cross more other orbits and expose themselves to interactions, including collisions, with many other larger bodies. Obviously, big targets are easier to hit than small ones; but there are several other important factors that enter in.

First, consider a small projectile on a course that will take it near a potential target. The target's gravity will attract the projectile, bending its path inward, a phenomenon called gravitational focusing. Every target, because of its gravity, will sweep an area larger than its actual cross-section area. The larger the mass of the target, the farther its reach. Double the radius of a body, and its mass (and volume) grows by a factor of eight: small bodies act a little bigger than they really are; but large bodies act *much* bigger than they really are. The mass-rich bodies get richer; the mass-poor bodies don't.

In fact, the situation is even worse for the little guys. Consider a swarm of bodies in which the largest are roughly the size of the Moon, with escape velocities of 2 or 3 kilometers per second. Perturbations of the smaller bodies by the largest will build up their random velocities toward the same value. Suppose two small bodies, each with an escape velocity of 0.1 kilometers per second and a randomly directed speed of about 2 kilometers per second, collide with each other. The most likely impact speed is, say, 3 kilometers per second. The two bodies smash each other to flinders in a violent explosion. Much of the collision energy goes into crushing rock. Rocky debris hurled out of the impact site at 0.5 to 1 kilometer per second easily escapes the combined gravity of the collision fragments, and both bodies cease to exist.

Now suppose one of these small bodies collides, not with another little guy, but with a lunar-size body. The impact will occur at a higher speed than for the two small bodies, because the gravitational acceleration of the bigger one will draw them together; let's say 4 kilometers per second. Again, most of the energy goes into crushing rock. The debris from the impact explosion will spread outward at an average speed of 1 to 1.5 kilometers per second, which is less than the escape velocity of the target. Very little mass will actually escape, and the target body will grow as a result of the collision. So the mass-rich bodies do indeed get richer, and the mass-poor ones actually get poorer. This process is called *runaway accretion.* It tends to produce planetary systems with a few massive rocky planets rather than hundreds of Moon-size planets.

During the T-Tauri phase, the size of the rocky bodies also matters. Every T-Tauri star gives off a dense solar wind of protons and electrons with a strong embedded magnetic field. The T-Tauri solar wind drops off in intensity the greater the distance from the Sun, at a rate that is somewhere between $1/R$ and $1/R^2$, where R is the distance from the Sun. At 2 AU from the Sun, the maximum rate of heating is somewhere between one-half and one-quarter of that at 1 AU. Here's how a T-Tauri star works to heat a solid body: the magnetic field embedded in its dense solar wind is dragged through each asteroid in its path. If the body is a good conductor of electricity (iron metal), the magnetic field will barely penetrate the surface at all. Like a dolphin sliding through a kelp forest, the asteroid will slip through the field lines and not get entangled in them. If the body is only a moderately good conductor (a carbonaceous asteroid), the magnetic field lines penetrate deep into it. The field lines are dragged through, inducing, as in an electric motor, a current that runs through the body at right angles to both the field lines and their direction of motion. The electric current heats the interior. Finally, if the body is a poor enough conductor (such as dry enstatite) the field lines simply slide through unimpeded. No current is induced, and no heating takes place.

COMPOSITION AND location are not the only important factors in solar-wind induction heating: size matters as well. Consider three asteroids, identical in composition (all with moderate conductivity) and at the same distance from the Sun, with diameters of five hundred, fifty, and five kilometers, respectively. Suppose that their conductivities are such that the T-Tauri solar wind can penetrate about twenty-five kilometers into them. The largest will be strongly heated at a depth of 25 km, and a layer at that depth will melt and differentiate according to density. A dense melt of metal and sulfides will sink to the bottom of the melted region, where it will rest upon the cold, undifferentiated interior. The melt then freezes to make a layer of metal and sulfides. This highly conductive metal layer will afford perfect protection against any further heating of the interior. The silicate component of the melt will rise because, once the metal and sulfide components are removed, it is less dense than the undifferentiated mixture that constitutes the surface layer. These lavas may erupt onto the surface and flood it. The middle-sized body, however, will begin to melt at its center, not at a shallow depth. Because it is so small and the melt is so close to the center (where the acceleration of gravity vanishes), the separation of its melts according to density may be very inefficient, giving the body more the structure of raisin bread than a well-layered mousse. The surface may remain unaltered. Finally, the smallest body, which is essentially transparent to the magnetic field, will be only gently heated, and remain unmelted and undifferentiated.

Small bodies have a further problem in that, once heated, they cool relatively quickly. (The rate of heat production inside them, and their total heat content, is proportional to their mass [i.e., to the cube of their radius], and their rate of cooling is proportional to their surface areas [i.e., the square of their radius]. The ability of a body to retain its internal heat for a long time is given by the ratio of its heat content to its rate of heat loss [i.e., proportional to its radius].) The smallest asteroids are so good at cooling themselves that it is nearly impossible to make them hot enough to melt. Asteroid-size bodies, which can conduct heat out of their interiors in millions of years, are in any case not much affected by long-lifetime energy sources such as radioactive decay of potassium, uranium, and thorium, which take billions of years to deliver their heat: the heat leaks out as it is generated and does not accumulate enough to permit melting.

These effects are not unique to solar-wind heating. There is another early heat source, radioactive decay of the isotope aluminum-26, that clearly occurred in the earliest days of the Solar System. The product of its decay, magnesium-26, has been found in very ancient aluminum-rich, magnesium-poor refractory mineral grains in meteorites; however, there is no clear evidence that aluminum-26 was abundant enough and widespread enough to be a major heat source in the Solar System. One problem is that the half-life of aluminum-26 is very short, only 720,000 years. Within the range of our uncertainty about how long it took the planets to accrete, most of the aluminum-26 may have still been present when the planets formed, or most of the aluminum-26 may have already been gone. If it had been present, though, it would have thoroughly melted the planets as they were accreting.

For any radioactive heat source strength, there is a critical size for bodies containing it below which the body cools rapidly enough to prevent melting, and above which the body melts. For the suspected abundance of aluminum-26, that critical size could be as low as a few kilometers. But many asteroids far larger than that size show no evidence of strong heating. This casts doubt on the general importance of aluminum-26 heating.

Size effects remain important for planet-size bodies, thousands of kilometers in diameter. Let us consider four planets, identical in composition and distance from the Sun, that have diameters of 3000 km (roughly lunar-size), 6000 km (roughly Mars-size), 12,000 km (similar to Earth or Venus), and 24,000 km (a rocky superplanet). Each one has a mass eight to ten times as large as its predecessor. The smallest, unless it has been heated strongly by the T-Tauri solar wind, short-lived radionuclide decay, or a catastrophic impact event, may have trouble melting under the influence of potassium, uranium, and thorium decay because it can cool itself at about the same rate that radioactive decay heats it. The Mars-size body will melt rather thoroughly, but it will cool rapidly after the first billion years or so, as the intensity of the radioactivity drops off due to decay. The interior should have limited opportunity for remelting, and release of volatile elements from its interior may be quite inefficient. The Earth-size body should retain heat well enough to keep it active for billions of years, with intense remelting, outgassing, and cycling of its crust and mantle. And finally, the superplanet should be even more vigorous in its activity than Earth for an even longer period of time. These ideas agree well with Mars and Earth as we know them, but we still lack a real rocky planet big enough to test our predictions of how superplanets ought to behave.

Suppose that the volatile materials (water, nitrogen, carbon dioxide, and so on) in these planets are released with efficiencies of 1 percent, 10 percent, 50 percent, and 90 percent, respectively. The actual mass of volatiles is the product of this efficiency times the total mass of volatiles in the planet: since we are comparing four bodies with identical compositions, the masses of total volatiles in these planets are, respectively, 1, 8, 80, and 1000. The relative amounts of volatiles released are then 0.01, 0.8, 40, and 900, respectively. The surface areas of these planets are in the proportion 1, 4, 16, 64. Therefore the amount of volatiles per unit area of the planet grows in the sequence 0.01, 0.2, 2.5, 14. Such a rapid growth of the atmospheric density with planetary radius suggests that even if the greenhouse effect is ineffectual on the smaller bodies, it may still cause powerful warming of the surfaces of the largest rocky planets.

Of course, a planet must melt and differentiate in order for it to release its internal volatiles to make atmospheres and oceans. Differentiation implies the formation of a core, a mantle, and a crust, as well as the hydrosphere and atmosphere. The core may seem sufficiently far removed from surface activities that it can be ignored, but we would do so at our peril. The core is the source of the planetary magnetic field, which helps protect the surface from energetic cosmic radiation and from interaction with the solar wind. The energy source that drives the generation of planetary magnetic fields has long been a matter of debate.

Whatever happens in the core, the melting and differentiation of an initially homogeneous planet causes extensive redistribution of the radioactive elements. Most of the uranium and thorium, and a major part of the potassium, end up in the crust, which makes up only about half of one percent of the mass of Earth. If all the radioactive heat sources were extracted into the crust, we would have trouble keeping the core molten and churning.

Heat coming out of Earth's deep interior has another very important effect. It drives mantle convection, which in turn drives continental drift, which then causes subduction of old crust into the mantle and generates new crust at midocean ridges. The engine that drives crustal recycling is motions in the mantle, which in many cases can be traced all the way down to the core-mantle boundary. Heat sources in the crust, no matter how large, are incapable of driving these deep-seated motions. The most appealing explanation is that the power comes from the core. Some of the heat surely comes from slow crystallization of the inner core, and some is surely due to dissipation of mechanical energy at the core-mantle boundary. (The precession rates of the core and mantle under the influence of lunar and solar tides are not equal, so they must "rub" against each other and produce heat.) But it is unlikely that these sources can provide all the energy that seems to be coming out of the core. A major radioactive heat source in the core would do the job nicely.

Evidence drawn from theory and recent laboratory experiments suggests that the heat source in the core may be radioactive decay of potassium-40. In one set of experiments, carried out by John G. Badding of Pennsylvania State University and his coworkers, potassium metal was found to dissolve readily in nickel (a major ingredient of the core) at pressures of three hundred thousand atmospheres, characteristic of the deep mantle. Another set of experiments, as yet unpublished, finds potassium from feldspars extracted into an iron-iron sulfide melt at high pressures. I have rather a fond attitude toward this idea because I first suggested the extraction of about half of Earth's potassium into the core in a paper published back in 1971.

The size of a planet also matters in several other important ways. For example, size strongly influences the acceleration of gravity at the planet's surface. The surface gravities for our sequence of four planets are 0.17, 0.37, 1.0, and 3.5 gravities. The pressure exerted by the atmosphere is simply its weight, which is the mass of the atmosphere multiplied by the acceleration of gravity. The relative atmospheric pressures on these planets should then be 0.0017, 0.074, 2.5, and 49, respectively.

Higher gravitational acceleration also tends to flatten out planetary topography. If the rock strengths are similar on the surfaces of all four bodies, the heights of the highest mountains should be in the proportion 6, 2.5, 1, 0.3. But interior temperatures tend to be higher in the larger bodies. High temperatures make the rock softer and less able to withstand the weight of high mountains. Thus the mountains on the superplanet are probably significantly less than one-fifth, and possibly less than one-tenth, the height they would have on the Earthlike planet.

Another important difference caused by differences in mass is the escape velocity of the planet. The escape velocity, which depends on the square root of the ratio of the planet's mass to its radius, is the speed that an object (spacecraft or atom) must have in order to escape completely from the gravitational field of the planet. The escape velocities for our four planets are then in the sequence 2.2, 4.4, 10.0, and 24.2 kilometers per second. These differences have a profound effect on the evolution of planetary atmospheres. A body with an escape velocity of 2.2 km/s cannot keep an atmosphere, because a significant fraction of the molecules in that atmosphere have thermal speeds that are larger than escape velocity: these gases escape readily and cannot accumulate to develop a significant atmospheric pressure. Therefore the smallest of our four bodies quickly loses even the feeble atmosphere that it is capable of generating. Water vapor is lost along with the atmosphere, stripping the planet of ice and water as well as air. The planet dies a very early death.

The Mars analogue, with an escape velocity of 4.4 kilometers per second, leads us naturally to the real Mars for an understanding of the stability of its atmosphere. As it happens, early spacecraft investigations of Mars were most enlightening because they revealed to us the existence of a new, previously unsuspected way that planets can lose atmospheres, a method that we did not and could not learn from studying Earth. Mars's trick is a little complicated, but well understood. Here's how it works. In the upper atmospheres of planets, extremely energetic photons (extreme ultraviolet sunlight) strike molecules and strip electrons from them, making ions. The ionized layer in planetary atmospheres is called the ionosphere. There molecules such as nitrogen (N_2) and carbon monoxide (CO) are made into the N_2^+ and CO^+ ions. The same sort of thing happens on every planet; the only difference is that different molecules are present on different planets. From time to time, a free electron encounters one of these positively charged ions and recombines with it (neutralizes it). That reaction dumps so much energy into the molecule so quickly that the molecule splits apart (dissociates) in a miniature explosion, firing two "hot" atoms in opposite directions at speeds of about 5 kilometers per second. Higher speeds are not possible because the strengths of real chemical bonds and the ionization energies of real molecules do not permit them. Therefore an upward-bound hot atom on a planet with an escape velocity of 4.4 km/s will be lost from the planet. On a planet with an escape velocity of 6 km/s, none of these hot atoms will escape. Mars-size bodies therefore get stripped of carbon, nitrogen, and oxygen atoms by this *dissociative recombination* process, while Earth- and Venus-size bodies are unaffected. Again we conclude that the mass of a planet matters greatly.

A further payoff from the study of Mars has been the realization that impacts may blast gases away at speeds above escape velocity. Hampton Watkins and I suggested in 1981 that "routine" comet and asteroid impacts could blast holes in the tenuous Martian atmosphere, accelerating the impactor debris and a portion of the atmosphere to escape velocity. More recently, Ann Vickery and Jay Melosh of the University of Arizona have carried out much more detailed calculations that show convincingly that *explosive blowoff* of atmosphere is very important for Mars. Over the age of the Solar System, as much as 99 percent of the Martian atmosphere may have been lost in this way. But this process is ineffectual for planets similar in size to Earth or Venus: the blast-wave speeds from impacts rarely reach the escape velocities of planets much larger than Mars. Erosion is therefore not very important. Impacts readily add to, and rarely deplete, their store of volatiles. Once again, the poor get poorer and the rich get fat by swallowing tasty comets and asteroids.

More of the same is not the same.

6

MERCURIES: TOO LITTLE TO COMPETE

MERCURY IS MORE THAN JUST a planet: it is an example of a type of planet that, from what we know, must be widespread in the Universe. How would we recognize another member of this class? First of all, it is usually the closest planet to its star. Second, because of its proximity to its star, it is extremely hot. Third, because its formation environment is asymmetrical, with lots of masses of solids available outside its orbit and little or no available mass inside its orbit, it is probably of relatively low mass. Fourth, for the same basic reason, the planet normally has a fairly eccentric orbit. Fifth, because it is formed closest to its star, it is probably (not definitely) the most refractory planet in the system. Therefore it probably has no present-day volatiles such as water, nitrogen, carbon oxides, and so on. Sixth, because it is refractory-rich, it has elevated concentrations of radioactive uranium and thorium. Seventh, it has been heavily bombarded and eroded, and is therefore probably deficient in silicates. Eighth, it is likely to be either despun or in a spin-orbit resonance. Ninth, because of its inability to hold an atmosphere and its very high temperature, life is very unlikely to have originated and persisted on the planet. These points are worth pursuing in greater detail, to gauge the variations that can occur within this general pattern.

FIRST, MERCURY-LIKE planets, or Mercuries, are in close proximity to their parent star. This appears to be almost a matter of definition of the class. But computer models of the growth of planets from vast swarms of smaller bodies allow individual examples of anomalous migration. Calculations done by George Wetherill of the Carnegie Institution in Washington, D.C., have shown that Mercury-like bodies (small, close, and eccentric) often occur in his computer simulations of the formation of the planets, but sometimes they are pathological cases. One of these actually originated beyond Mars and worked its way inward, surviving the danger of catastrophic collisions with the growing planets only to end up in the orbit closest to the Sun. Such a vagrant, once settled down close to its star, must then be subjected to all the rigors of life close to a star for the rest of its lifetime. Think of it as a particularly unpalatable form of early retirement.

Second, Mercuries are hot. This is an unavoidable consequence of close proximity to a star. How hot is hot? Hot enough in the daytime to melt lead or aluminum (although it is enormously improbable that either metal would exist on the surface: lead enters the core, and aluminum is tied up in refractory silicate minerals). Any true Mercury (one that originated close to the Sun) started out hot and continues to be baked. The equator, of course, gets the brunt of it. Conditions do not improve with time. Over the course of the main-sequence lifetime of the central star, the luminosity of the star edges upward. When the star finally depletes its hydrogen fuel it may, if it is massive enough (about 60 percent of the mass of the Sun or larger), switch over to helium fusion in the center of the helium core. This change in nuclear behavior heats and expands the core of the star, which lowers the acceleration of gravity for the entire outer envelope of the star, causing it to expand greatly in radius, sometimes by a factor of hundreds to thousands. The helium-burning star will normally expand enough to swallow up its innermost planets, becoming a full-fledged giant. Mercuries are the first to go.

Third, Mercuries have low mass. One of the consequences of formation at the inner edge of the system is that there was no significant condensed material closer to the central star: almost nothing was condensed at the very high temperatures prevalent in that inner region. During the era of collision and growth of asteroid- and lunar-size bodies, far less mass could have been collected. The orbital periods of bodies this close to the center of the system are short, and the relative periods of bodies in close orbits are short, so the planet probably accreted to its ultimate size rather quickly. But once the local supply of solids ran low, the only bodies that could strike our little Mercury were those propelled inward by larger, more distant planets (the "Venus" and "Earth," and later the "Jupiter" and "Saturn" of that system). These projectiles arrived on more highly eccentric orbits, approached our Mercury at high speeds, and impacted very violently. These late-arriving materials, originating much farther from the Sun, strike with so much energy that they are largely blasted back off into space, carrying with them debris from the surface of the planet. In short, the later stages of planetary accretion are, for the little guys, a time of violent erosion. Growing, and accreting volatile-rich material, are out of the question: Mercuries are too little to compete at the feed trough. They are the runts of the stellar litter—the older they are, the runtier they get.

Fourth, Mercuries have eccentric orbits. Another important consequence of forming at the inner edge of the dust disk is that the available supply of solids is asymmetrical. There is less opportunity for accretion to average out the orbital motions of all the fragments, and a much greater likelihood of making a planet with a significantly elongated, eccentric orbit. The word *eccentric* is actually rather descriptive: planets follow elliptical, not circular, orbits.

It is easy to construct an ellipse by sticking two thumbtacks into a soft surface and tying the ends of a string to them. Generally the string is longer than the distance between the two tacks, so there is some slack in the string. Now use the point of a pencil to pull the string taut. Put the point of the pencil straight down on the paper and slide it along the string until the pencil point reaches its closest approach to one of the pins. Then put the string on the other side of the pins and repeat the action. The resulting curve is an ellipse. It is, specifically, the locus of all points, the sum of whose distances from two fixed points (the pins) is equal. The fixed points are called the focal points or foci of the ellipse. In planetary, satellite, and stellar motions, the "central body" (Sun or planet) sits at one of the two focal points, called the prime focus. The other (secondary) focus is empty. The planet gains energy (speed) as it falls closer to the central star, and slows as it coasts outward to the opposite end of the ellipse. Even if no central star were visible, there could be no question which was the prime focus: the motions of the planet would give it away unambiguously.

The long axis of the ellipse is called the major axis; half its length is called the semimajor axis. The semimajor axis is also mean distance of the ellipse from the prime focus. For a planet, it is the mean distance of the planet from the center of its star. Note that, if the two pins are very close together or the string is long, the resulting ellipse is very close to a circle. The prime focus of a circle is of course the center of the circle. But the central body in an elliptical orbit is at the prime focus, not the center, of the ellipse. The motion of its planet is eccentric, meaning that the planet does not move symmetrically around the center of the ellipse. The eccentricity, e, which describes the shape of an ellipse, is $e = 0$ for a circle and $e = 1$ for an infinitely elongated ellipse (a parabola). Popular understanding notwithstanding, orbits are neither "oval" nor "egg-shaped." The egg is, first of all, a three-dimensional body, not a two-dimensional line on a plane. Eggs are distinctly asymmetrical, with one end always fatter than the other. Ellipses are symmetrical. As for *oval*, the word means "egg-shaped." Aficionados of both biped and quadruped racing will notice that the word *oval* is commonly misapplied to a pair of parallel lines capped on both ends by semicircles. This is yet another example of eccentricity.

Because the planet moves faster when close to its star, and because so little of the orbit is close to the star, the planet travels at a high angular rate as seen from the star when it is near the point of closest approach. The word we use for the point of closest approach of Solar System bodies to the Sun, *perihelion*, literally means "close to the Sun." For bodies orbiting a star, we refer to the point of closest approach as the *periastron*, "close to the star." The point of greatest distance from a star is called the *apoastron*, by analogy with the *aphelion* of a body orbiting the Sun. The periastron distance may be much less than the apoastron: for Mercury, the perihelion is at 0.307 AU from the Sun and the aphelion ("far from the Sun") is at 0.467 AU. The eccentricity of Mercury's orbit is 0.206, and its mean distance from the center of the Sun is 0.387 AU. (It is actually very easy to calculate these numbers from each other: perihelion distance is the semimajor axis times $(1-e)$, and the aphelion distance is the semimajor axis times $(1+e)$. The sum of the perihelion and aphelion distances is exactly twice the length of the semimajor axis.)

High eccentricity means that the planet varies greatly in its orbital speed and its distance from its sun. It also means a large variation in the intensity of sunlight falling on the planet, and therefore a large difference in temperature as it moves around its orbit.

Fifth, Mercuries are volatile-poor. At the time of formation, water and gases are virtually absent in the planet, which is mainly composed of volatile-poor materials that are stable at high temperatures, deep in the interior of a prestellar nebular cloud. The surface environment of such a planet, lacking the protective and moderating effects of an atmosphere, is extremely harsh.

The prospects for a true Mercury to pick up volatiles later in its career are distinctly limited. There are, it is true, many impacts of volatile-rich comets and asteroids over the lifetime of such a planet. Unfortunately, the comets strike at enormous speeds, sometimes exceeding one hundred kilometers per second. The blast waves from their impact explosions easily escape from the planet, carrying along Mercury surface material. An occasional water-bearing asteroid on a much less eccentric orbit, perhaps an extinct short-period comet, can impact at low speeds, perhaps as low as five to ten kilometers per second. The vapor from such an impact has about one hundred times less energy than the more extreme comet impacts. Also, the vapor from the asteroid explosion is made of much heavier molecules than the hydrogen-rich vapor from comets, and therefore travels at even lower speeds. Some fraction of that vapor does not immediately achieve escape velocity, but instead follows a long, arching ballistic ("like a thrown stone") trajectory that strikes the surface of Mercury somewhere remote from the impact site. Most of this vapor hits a surface that is either very hot or will be very hot as soon as the next sunrise comes around: the vapor that fails to stick on the surface is soon ionized by ultraviolet radiation from the star and swept away by the star's stellar wind. A small fraction of the vapor, however, probably about 0.1 percent of it, is fortunate enough to land on the permanently shadowed floor of a crater close to one of the planet's poles. There it freezes to the incredibly cold surface (perhaps less than one hundred degrees above absolute zero) and forms a deposit of ice.

With the sole exception of these small polar ice deposits, Mercuries are devoid of volatiles. Liquid water is completely impossible; even warming the ice in some manner would cause the ice to sublimate (a word that denotes the direct evaporation of a solid without melting) at temperatures well below the melting point. Any substantial atmosphere is impossible, because escape is just too easy.

Wandering planets, those that didn't start out as Mercuries, will be baked by the nearby star and release gases into their atmospheres. But the intense ultraviolet radiation of the nearby star will tear apart the molecules in the atmosphere, heat the upper atmosphere to scorching temperatures, and fire hot atoms about at speeds of several kilometers per second. Any surface minerals that bear water of hydration, or carbonates, will decompose and release their gases, which promptly feed into the disassembly line in the sky.

Sixth, Mercuries are refractory. They are formed with high concentrations of all refractory elements, including calcium, aluminum, and titanium. Accompanying these elements are the very rare but very important refractory radioactive elements, uranium and thorium. How much of these elements ends up in the planet is an open question. Suppose the planet starts out highly refractory, forming at temperatures so high that the magnesium silicates are not completely condensed and accreted. Then the planet must have a high concentration of radioactive elements, and has a very good chance of melting and differentiating. Other heat sources help out in abetting early melting, such as the fact that the material starts out hot because it is so close to its star, and the gravitational potential energy released by the accretion of its material liberates enough heat to warm the planet a thousand degrees or so. Also, it is very strongly heated by the T-Tauri-phase stellar wind, and it can be heated strongly by tidal interactions with the star. Short-lived radioactive isotopes, such as aluminum-26, may also contribute.

Fine; there are lots of heat sources. Let's say the planet started melting very early in its history. Then a dense metal melt sinks (converting more gravitational energy into heat as it trickles downward to form the core) and a crust of calcium aluminum silicates (anorthite) is sweated out of the mantle along with a wide variety of incompatible elements, including uranium and thorium. As the planet is bombarded by comets and asteroids, the part struck and eroded (the crust) is precisely the part where most of the uranium and thorium reside. Of course, these heat-producing elements would not be so exposed, and would still be uniformly mixed in throughout the planet, if differentiation had not yet occurred. Thus one direct consequence of differentiation is that the incompatible refractory elements will be preferentially exposed at the surface and then preferentially lost.

Seventh, Mercuries are heavily eroded by cosmic bombardment. Late, high-energy impacts are very effective at blasting off the outer layers of the planet. Most of the fine dust and rocks kicked off Mercury will slowly spiral into the Sun, although some may be perturbed by Mercury into planet-crossing orbits and end up falling as meteorites on Venus, Earth, or Mars. Certain classes of meteorites have from time to time been suggested as possible Mercury surface material. Mercury, after erosion of about 70 percent of its silicate rocks, would be left as a silicate-poor planet. The post-erosion residue is about 60 percent metal-rich core material, compared with about 30 percent core in Venus and Earth.

Eighth, Mercuries are readily captured into orbital resonances. The eccentricity of a planet's orbit causes the gravitational force exerted on it by its star to change as it follows its orbital path around the star. A planet that is quite close to a star is distorted by the star's gravity, elongated so as to produce a bulge on the side toward the star and, by symmetry, a bulge on the opposite side as well. As the planet nears periastron, the force exerted by the star raises a larger bulge, and the gravitational attraction of the star for that bulge grows even more rapidly than its attraction for the planet as a whole. The attraction of the star for the tidal bulge tends to oppose the rotation of the planet. Since the planet is a very viscous body, mostly solid, the rising and falling of the tidal bulge as the planet follows its eccentric orbit deposits energy in the planet, just as flexing a paperclip causes it to heat up. The heat comes from the eccentricity of the orbit and the rotation of the planet. Sometimes so much rotational energy is lost that the planet loses its spin and becomes rotationally locked on to its star: it always keeps the same side pointed at the star. There are many examples of this kind of behavior in the Solar System, in which the spin period and orbital period are equal. When the spin and orbit periods are harmonically related (in a ratio of two small whole numbers), we say that there is a resonant relationship between the two. The most familiar example of a one-to-one spin-orbit resonance is the Moon, which always keeps the same side directed toward Earth. The size of the tidal bulge in the Moon, which has been measured both directly and indirectly, is about 2 km. The four large Galilean satellites of Jupiter are also locked in a one-to-one (1:1) spin-orbit resonance with Jupiter.

When the raising of the tidal bulge is considered, the tidal force exerted by a body, such as a star, is proportional to its mass and inversely proportional to the *cube* of the distance between it and its planet. It is clear, then, that a planet like Mercury, which varies in its distance from the Sun by about a factor of 1.5, must experience much stronger tidal forces near perihelion; in fact, the perihelion tidal force is larger than that at aphelion by a factor of $1.5 \times 1.5 \times 1.5$, or 3.37. The orientation and motion of the tidal bulge when Mercury is near perihelion is therefore *much* more important than what the bulge is doing when the planet is near aphelion.

The result of this strong dependence of tidal force on distance is that the angular rate of rotation of Mercury is just that needed to point the tidal bulge at Mercury during the time of perihelion passage. Note that the angular rate of rotation of Mercury is constant, 360 degrees in 58.646 Earth days, or 6.1385 degrees per day. The *average* rate of progress of Mercury around its orbit is 360 degrees per orbital period of 87.9686 days, or 4.0924 degrees per day. But the *actual* angular rate of Mercury's motion along its orbit varies substantially with position on the orbit, being much higher near perihelion; in fact, it matches the spin rate! The real Mercury, as was discovered by MIT radio astronomer Gordon Pettengill and coworkers, actually executes 1.5 revolutions per orbit (at successive perihelion passages, opposite tidal bulges point at the Sun). The situation repeats exactly after two orbits (three rotations), corresponding to a 3:2 spin-orbit resonance.

If the tidal forces were much stronger, or prolonged for a much longer time, the eccentricity of the orbit would slowly damp out. Some of the orbital energy of the planet would be converted into heat, keeping the interior of Mercury hot.

Before the MIT radar observations were made, it was long believed that Mercury always kept the same face toward the Sun, baked eternally on one side, with the other side freezing in perpetual darkness. But we now know that Mercury rotates: to paraphrase James Bond, Mercury is "broiled, not baked."

Ninth, Mercuries support no life — ever. A planet that forms with no volatiles, is scorched by temperatures high enough to melt lead, never has any significant atmosphere or any oceans, loses atmosphere rapidly, and lacks even a normal crust, is not a plausible home for life of any imaginable sort.

The picture is a dismal one. Mercury, and all its cohorts in deep space, have fallen catastrophically off the rails of planetary evolution.

GIANT IMPACTS: CHIPS AND BROKEN PLANETS

PLANETS, WHEN THEY ARE YOUNG, spend a lot of time playing on the freeway. Planetary accretion necessarily involves disturbing the orbits of asteroid- and lunar-size bodies up to moderate eccentricities. As the Russian theorist V. S. Safronov pointed out nearly a half-century ago, a swarm of partly grown planets will be stirred up by the gravitational disturbances of the bodies in the swarm to speeds roughly equal to the escape velocities of the largest members. Consider several hundred roughly lunar-size bodies with two or three Mars-size bodies at the head of the class. The escape velocities of the biggest bodies are about four or five kilometers per second. Near-collisions are frequent. A flyby within a distance of ten planetary radii (a strongly perturbed near miss) is one hundred times more common than an encounter within one planetary radius (a collision).

Anyone who has ever driven through the Atlanta area has been impressed by the instant, unsignaled high-speed lane changes executed by small, strongly redshifted automobiles (and pickup trucks). About one in one hundred of these dazzling maneuvers results in a catastrophic collision that destroys the smaller vehicle. It is humbling to realize that this sort of thing was going on in the Solar System long before the rise of intelligent life.

Now, what happens to a Moon-size body that flies by a Mars-size planetesimal at close range, narrowly avoids a collision, and sails off into the sunset? The size of the kick depends on how close the bodies get, and the direction of the kick depends on whether the smaller body flies by ahead of, behind, above, or below the larger. The gravitational kick imparted by the bigger body will change the speed of the smaller body randomly by one to five kilometers per second, enough for a substantial lane change. Now suppose that Mars-size body was near the orbit of Venus; in fact, suppose it was Venus itself in an early stage of growth, with only a tenth of its ultimate mass. The orbital velocity of Venus on its almost perfectly circular orbit was about thirty-five kilometers per second. A lunar-size body starting in a nearby near-circular orbit can, after a five-kilometer-per-second kick, find itself in an orbit that reaches in as close as 0.385 AU from the Sun (Mercury, in the inner lane, is at 0.387 AU) or as far out as 1.258 AU (Earth, out in the third lane, is defined to be at 1.000 AU). Clearly, a moon-size body can be accelerated by a single close encounter with a proto-Venus to speeds high enough to take it in to cross Mercury's orbit, or out to cross Earth's orbit, making a collision with either planet possible. And of course every such orbit still intersects the orbit of Venus, so that yet another orbital disturbance by—or a collision with—Venus is possible. Later in the accretion process, when the largest bodies are nearly full-grown, even Mars-size bodies can be kicked about so that they can cross the orbits of several planets.

This is, if you think about it, a fairly hair-raising prospect. We have seen in recent years what a ten-kilometer projectile did to Earth at the end of the Cretaceous era, when the dinosaurs expired. That was a 250-million-megaton blast, roughly equivalent to twenty thousand World War IIIs happening within one hour. No wonder it was so lethal! But what about a Mars-size (6000-km) body colliding with Earth? A line from a Cole Porter song in the movie *High Society* casually predicted that a collision between Mars and Earth would occur "next July." The film was made in 1956: the song's collision prediction has turned out to be no more reliable than those in supermarket tabloids.

Failing to find in the traditional literature a reliable account of a Mars-Earth collision, we are forced to turn to supercomputer calculations, such as those done by Jay Melosh of the University of Arizona. This is the safest way to find out what happens in such a collision, and it is also much cheaper than carrying out even a half-scale experiment.

Here's the magnitude of the problem: a 6000-km body will deliver about two hundred million times as much energy as the Cretaceous impact. That's thirty thousand times as much energy as would be needed to boil away all the oceans and sterilize Earth's surface. It would ruin your whole eon.

According to Melosh, the outcome of a giant impact depends rather sensitively on whether it is a perfect head-on collision, or a bit off-center, or a grazing impact. He finds a range of conditions in which the off-center impact of a Mars-like planet splashes a tremendous spray of molten rock out ahead of the impact point, part of it traveling fast enough to orbit Earth but not fast enough to escape from Earth's gravity. The core of the Mars-size body hits Earth squarely and is stopped, contributing almost nothing to the spray of ejected silicates. It then sinks into Earth's core. About half of Earth's crust is utterly demolished by the impact. Part of the crust and upper mantle is ejected, largely vaporized, into the plume of liquid rock and vapor that is hurled into space. Much of the splash of molten rock falls back to Earth, striking everywhere. The impact adds enough mass to increase the radius of Earth by nearly two hundred kilometers. The atmosphere is heated to thousands of degrees, the oceans boil away, and any primitive life that has arisen is destroyed.

The plume of molten rock and rock vapor that has been hurled into orbit around Earth quickly loses all its volatiles. Even the more volatile rock-forming elements, such as sodium, sulfur, and potassium, are largely lost. Within a few hours the white-hot debris cloud forms a broad ring around Earth. As the ring radiates off heat it cools and solidifies into a vast swarm of solid particles, which quickly collide and begin to accrete into larger bodies. Under the impetus of tidal forces, the growing satellite retreats steadily from Earth, sweeping up material as it goes. Earth cools, enormously thick layers of cloud form, and the oceans rain back onto the molten, ravaged Earth for millions of years, gradually cooling it until the crust is solid. Finally, the crust cools enough for liquid water to collect, and the oceans begin to reaccumulate. The satellite, still close to Earth, raises towering tides, while the tidal interactions continue to accelerate it outward. Life begins, develops as the skies clear and the satellite retreats, and eventually evolves an intelligent species that calls the new satellite "the Moon."

From what we understand of the origin of planets, collisions of bodies that are at least as large as the Moon must be an important part of the growth of every rocky planet. Mars-size impacts probably occur about once per Earth-size planet, on average. But that does not mean that the outcome of the collision must be the same in every case. On the contrary, many of the impacts must result in different but equally spectacular results, such as despinning or reversing the spin of the planet (Venus comes to mind), excavating and ejecting its crust and upper mantle (as on Mercury), or tipping its spin axis (Earth, Mars, Saturn, Uranus, and Pluto). Shallow grazing impacts may eject a flat disk from which multiple satellites may form, or spray highly evolved planetary material, including pieces of planetary crust, throughout the region of the inner planets.

The satellites made in this manner may be large, like the Moon. But much more frequent, smaller impacts may also eject enough material to make small satellites. All the satellites made by impacts of rocky planets should be thoroughly baked and very thoroughly purged of volatiles. These satellites would have no opportunity to develop atmospheres or oceans. Even cometary and asteroidal impacts of the sort that delivered water to Mercury's poles will have no more profound effect on such moons: indeed, the Clementine and Lunar Prospector space missions have returned strong evidence of the presence of ice in permanently shadowed crater bottoms near at least one of the lunar poles. But liquid water is and always has been unknown on the Moon. The lunar rocks are dry to a degree almost unachievable in laboratories on Earth. No sign of weathering by water or atmospheric gases has ever been found in returned lunar samples or in meteorites of lunar origin.

The Moon's desolation is a direct consequence of its fiery origin. The forested hills and dappled vales seen by Verne's space-bullet passengers (in *From the Earth to the Moon* and *Around the Moon*) not only do not exist—they never did.

SO MUCH of what we take for granted on Earth was shaped by the impact event that made the Moon. Earth's spin, axial tilt, and internal composition were all influenced, as were the Moon's composition and orbit and the tidal forces that cause the Moon to continue to evolve outward from Earth and decelerate Earth's spin. We are almost forced to ask, "What would Earth be like if the impact hadn't happened?"

The answer must to a very large extent be based on theory. We could imagine an Earth with more crust, or more atmosphere, or more water, but we have no direct evidence of conditions on Earth before that impact to test our theories. It would be convenient if we could examine another Earth that had not passed through such a giant impact and compare it point-by-point to our own planet; but that is unfortunately impossible. Even detecting Earth-size planets around other stars will remain impossible for many years to come, and studying their properties will be an even more enormous challenge. We shall return to this matter in Chapter 18.

8

A SUITE OF EARTHS

MORE OF THE SAME IS not the same. If Earth were bigger or smaller, it would not be the same Earth. But would Earth be more suitable for life if it were a little smaller? A little larger? How would the environment on Earth's surface change if the planet were a different size? What if its composition or its distance from the Sun were modestly altered? To revive the question asked by Plotinus, Ockham, and many others, what is the best of all possible Earths?

Let's suppose that we make up a big batch of Earth-stuff in our planet factory. All of it has exactly the same composition, with about 0.05 percent water, a couple of parts per million of nitrogen, 30 percent iron-nickel metal and sulfides, and so on. The mix is uniform, and exactly the same as Earth. Let the mixture simmer in a medium nebula for a million years. Now let's pour out this Earth-formula mixture into several molds of different sizes. The stuff warms up and melts, the core material sinks to the center, and the volatiles rise to the top. So they all turn out the same, right? Wrong! More of the same is *not* the same!

In Chapter 5 we talked about why the fate of a planet depends on its size. Now we take that theory into the laboratory for testing.

Consider a Mars-size Earthlet. Little Earthlet has one-tenth the mass, one-eighth the volume, and one-quarter the surface area of Earth. By the rules of the game, it has exactly the same composition as Earth. If Earthlet releases its water as efficiently as Earth, it would have one-tenth as much water on one-quarter as much surface area: the average depth of water on the surface would be only 40 percent of what it is on Earth.

Earthlet would differ in many other ways, as well. For example, the acceleration of gravity at its surface would be about 40 percent that at the surface of Earth. Falling objects on Earthlet, instead of picking up ten meters per second (thirty-two feet per second) of speed with each second of free fall, would pick up only four meters per second (thirteen feet per second). The lower gravity means that everything would weigh only 40 percent as much on Earthlet as on Earth. The pressure at the bottom of the "ocean" would be much less than on Earth. Not only is the water layer just 40 percent as deep; a given depth of water weighs 40 percent of what it does on Earth. The average pressure on the *abyssal plains* beneath Earth's oceans is about 320 atmospheres, so the average pressure at the bottom of Earthlet's seas is only 50 atmospheres. That same pressure is reached on Earth at a depth of just five hundred meters (about sixteen hundred feet).

The lower gravity also means that mountains weigh 40 percent less than on Earth. In principle, it should be possible to pile the rocks up two and a half times as great a height on Earthlet before they exert the same force on their base. The significance of this effect can be illustrated by a naive, intentionally oversimplified example. Suppose a typical crustal rock on Earth has a crushing strength of about 2000 atmospheres. (Most sedimentary rocks crush at pressures of about 1000 atmospheres; some igneous rocks remain intact at 3000 atmospheres' pressure.) For comparison, the tallest mountains on Earth, the Himalayas, reach a maximum height of about 9 km (5.4 miles). The weight of these high peaks upon a surface at sea level is easily calculated by multiplying the density of the rock times the height of the mountain times the acceleration of gravity: it is 2400 atmospheres. (Of course, it isn't all that obvious that sea level is the right reference level, but we are accustomed to using it.) On Earthlet, with gravity only 40 percent as strong, mountains of the same materials and temperatures could in principle be piled up to the enormous height of 22.5 kilometers (13.5 miles; 71,000 feet).

Sea level is a rather silly point of reference on planets with no seas: on Venus, for example, the average height of the lowlands is sensibly taken as the reference point. But on Earth, the real lowlands are the abyssal plains of the ocean floor. If we preferred to measure the height of the highest peaks relative to the most common ground level on Earth, the abyssal plains (3000 meters or 1.8 miles below sea level), then we would estimate a rock strength of 3200 atmospheres, which is about the limit for silicate rocks. Then the greatest elevations on Mars should be around 30 kilometers (18 miles).

It is difficult to verify this rough estimate experimentally, since we have no Mars-size planets of Earthlike material sitting around for inspection and experimentation. But we do have Mars. Mars, formed farther from the Sun, is intrinsically less dense, and presumably formed both more oxidized and more volatile-rich than Earth, both of which suggest that its material may be slightly weaker than Earth rocks. Nonetheless, let's give the comparison a try. The tallest mountain on Mars, Olympus Mons, reaches to a towering twenty-seven kilometers above the low plains. Now, the low plains are not "sea level." Perhaps they correspond more closely to the abyssal plains on the floor of Earth's great oceans. In any case, the height of the highest peak on Mars is well within the range expected based on an assumed similarity to Earth. Naive arguments are not always this reliable, but they are usually easy to apply!

Of course, if we were considering a planet that is larger than Earth, the mountains would be lower and the oceans deeper. Indeed, a planet modestly larger than Earth could easily have enough water to swamp its continental platforms, or even inundate the entire world! This tells us that having a reasonably good idea of how mountain heights vary on worlds of different sizes can be very important. Fortunately, we can easily think of ways to improve our estimate of mountain heights. Other factors that influence topography include the internal temperature of the planet, the rock type, its density and water content, and the dynamical activity of the planet. Mars, for example, has a colder surface than Earth because of its greater distance from the Sun. But Mars also, because it has ten times less mass (and ten times less radioactive heat sources) than Earth, but four times less surface area, has an internal heat loss rate (watts per square meter) that is about 40 percent that of Earth. This lower heat flow is carried through the crust by either conduction or by mass movement of magma (volcanic eruption). Conduction depends on the fact that the temperature in the crust of Mars, like that of Earth, increases with depth. The internal heat naturally flows from hot to cold—that is, outward. If conduction were the only way the heat could get out, the temperature gradient in Mars's crust, which is the rate at which temperature increases with depth, would also be about 40 percent as large in Mars as in Earth. If some or most of the heat is erupted rather than conducted upward, then the temperature gradient must be even smaller. But either way, rocks will get hot enough to soften and lose their strength at a depth that is at least two and a half times as deep on Mars as on Earth. The cold, rigid, brittle outer portion of a rocky planet, the *lithosphere*, generally does not correspond precisely to the chemically defined crust. Earthquakes can occur only in the lithosphere because only there are the rocks hard and brittle enough to fail catastrophically under stress. On Earth, the lithosphere extends through the crust and down into the upper mantle. The hotter, weaker region below the lithosphere, called the *asthenosphere,* flows under stress and cannot support loads. In effect, the solid lithosphere floats on the fluid asthenosphere. Like a cork or ice cube floating in water, the part that sticks up into the air is supported by a part that sticks down into the fluid. The buoyancy of the thick "roots" of mountain ranges supports the towering edifice of the peaks above them.

Now let's get back to Earthlet. For lack of better information, we would expect its lithosphere to be about two and a half times as thick as that on Earth. Then it would be reasonable for mountains to be two and a half times as high as those on Earth. Once again, we expect vertical relief on Earthlet to reach up to twenty-five to thirty kilometers, which, as we have seen, is demonstrably true of Mars.

Earth has, in addition to mountains, a complex system of extremely deep trenches in the ocean floor, some of them plunging more than nine kilometers (thirty thousand feet) below sea level. These trenches are maintained by the dynamic movements of Earth's crust. The trenches are produced by *continental drift*; more specifically, the motions of the continents require that crust be destroyed in some places and that new crust be made in other places. The thinnest, most vulnerable part of Earth's crust is the ocean floor, where almost all crustal destruction and creation take place. The thin, dense oceanic crust, when plowed along by plate motions, is sometimes forced under the edge of a thick, buoyant continental block. Long stretches of the continental margin, as for example along the eastern coastline of Asia, are paralleled by such *subduction zones*, in which the cold, thin, dense, and brittle oceanic crust bends downward and is forced under the edge of the continent, breaking and generating powerful earthquakes as it deforms. The descending oceanic crust underrides the deep roots of the continental mountains, eventually being driven to such depths in the hot mantle that it melts. The ocean floor crust carries a burden of sediments, rich in water-bearing clays, carbonate shells of marine organisms, and organic matter, all of which are decomposed and devolatilized as the descending plate begins to melt. The released gases, largely water vapor and carbon dioxide, form bubbles in the magma and lower its density, helping it to erupt in violent, gas-driven volcanic activity. Indeed, on Earth we see a ring of volcanoes completely girdling the Pacific Basin inland from the subduction zones, a ring of fire that puts the wildest fantasies of Johnny Cash to shame.

During subduction, the total surface area of Earth and the total area of crust must remain unchanged. If one region of the surface, such as an ocean basin, is being compressed, then another region must be stretched. In the areas where tension pulls the oceanic crust apart, magma wells upward from the mantle and freezes, upon contact with seawater, into fresh crust. The rising magma forms a long volcanic ridge on the ocean floor, straddling the crack. Further drift pulls the sides of the crack apart, and new magma wells up in the middle. This symmetrical, double-peaked system of ocean-floor volcanic chains is called a midocean ridge system, of which the best-known example is the mid-Atlantic Ridge.

But what happens on Earthlet, with its far thicker, colder, rigid lithosphere? Probably there will be no significant subduction and no ridge systems: if you slide two thin sheets of paper together, one will readily slip beneath the other — but if you shove two thick encyclopedias together, you get an impasse. That means no subduction zones, and therefore no deep trenches. Interestingly, this story seems to work well for Mars: no clear evidence of trenches, subduction zones, or "midocean" (lowlands) ridge systems can be seen. Mars does, however, have a chain of titanic volcanoes, some very ancient, of which only one, Olympus Mons, might conceivably still be active. The volcanoes constitute the Tharsis bulge. So enormous are these volcanoes that it takes prodigious amounts of magma to build them. It appears that all the magma generated on the entire planet had been squeezed out along a single line, building one gigantic volcano at a time. This is perfectly reasonable for a planet that has a very thick lithosphere — and that's exactly what we expect for a planet of this size.

Similar processes occur on Earth on a much smaller scale. Some volcanic chains, such as the Hawaiian Islands and the Emperor seamounts of the northwestern Pacific, show that continuing injection of magma from below can produce a long series of volcanic peaks in strict age sequence. It is as if a blowtorch of hot, buoyant magma was directed at the base of the oceanic crust, melting through to make a volcano, which then is slowly displaced by continental drift until it no longer sits above the hot spot and is no longer close enough for magma to find its way to the established vent. A new volcanic vent is then melted through the lithosphere above the hot spot, the magma begins to erupt there, and the previous volcano goes extinct. Based on our experience on Earth, identifying Mars's Tharsis Ridge volcanoes as magma extrusion from a crack seems less likely than a single hot spot in the Martian mantle, a plume of hot, rising magma, has slowly worked its way along the base of the Martian lithosphere, sequentially generating the giant volcanoes. That suggests that Mars's crust does in fact move—but on Mars it moves as a single piece, not as dozens of continental "rafts," as on Earth. This phenomenon, coupled with the evident dominance of one or two hot spots over billions of years, leads to a picture of a single-plume Martian mantle and single-plate drift.

Another consequence of a thick, immobile lithosphere with no continental drift is that the topography must be strongly shaped by large impacts as well as volcanism. The crust is so rigid that large impact scars remain readily recognizable even after billions of years. Indeed, about half of the surface area of Mars is dominated by heavy cratering. The surface is at first glance shockingly Moonlike, but closer examination shows the softening effects of erosion by the atmosphere and by ancient precipitation and runoff.

Now we return again to Earthlet and provide it with the general geology of Mars, supplemented by its complement of liquid water. Earthlet, like Mars, has no real ocean basins swept open by continental drift. The surface is much more rugged than Earth's because of the lower gravity and thicker lithosphere. It also has a much thinner average water layer, for reasons we have already mentioned. The water could easily pool in a number of deep, isolated seas, many in ancient impact craters.

A thick, cool crust and lithosphere have other implications. Reactive, generally acidic, gases vented from the giant volcanoes cannot be recycled by subduction. Instead, they chemically attack (weather) the crust to make clays, carbonates, sulfates, chlorides, and a variety of other minerals that accumulate as thick layers of nonrecyclable sediments. The seas must become saline, and sediments must incorporate volatiles irreversibly. The planet is so small that it lacks a mechanism for self-rejuvenation. As its internal heat source slowly fades away, its volcanic activity must also choke off, depriving the surface of a fresh supply of volatiles. The planet dies slowly, a prolonged death lasting billions of years, as crustal minerals slowly consume the water and other volatiles. If life has arisen on this planet, it must adapt to the deteriorating conditions or suffer extinction. It must find and use the most congenial environments on the planet, the hot springs and brackish puddles, the last refugia of life.

There is yet another consequence of Earthlet's small size: its escape velocity is about 4 kilometers per second, compared with 11.2 kilometers per second for Earth. Light molecules can escape from Earthlet far more readily than from Earth. But even worse, there are other mechanisms for loss of volatiles available to Earthlet (or Mars) that are completely ineffectual on a planet as large as Earth, but could strip a small planet of its atmosphere and oceans. The principal culprits are explosive ejection of gases by impact events and dissociative recombination, which was discussed in Chapter 5.

Impact erosion of the atmosphere is rather efficient for bodies as small as Earthlet. Even an early abundance of gases is no guarantee of a rosy present, let alone a survivable future. Impacts remove atmosphere unselectively: whatever is above the impact point has a good chance of being blown away. The dissociative recombination process is, by comparison, quite selective. It can eject nitrogen, oxygen, and carbon atoms, but is incapable of making hot atoms of the inert gases, and very inefficient at kicking them to escape velocity even indirectly, via collisions of hot carbon, nitrogen, and oxygen atoms with helium. The heavy rare gases neon, argon, krypton, and xenon, cannot be ejected by hot-atom collisions because they are too massive. Loss of atoms of carbon, nitrogen, and oxygen deplete the planet's supply of carbon dioxide and molecular nitrogen, leaving the atmosphere richer in the heavy rare gases.

Loss of atmosphere is not terribly sensitive to distance from the Sun, although there are some connections: the rate of loss by hot-atom ejection depends on the rate at which the central star supplies high-energy (ultraviolet) light, with enough energy to ionize molecules. That supply is very sensitive to the temperature of the star. There are vastly more high-energy photons in the light from hot O, B, and A stars than from cooler stars. Also, of course, the supply rate drops off with the square of the distance between the planet and the star. Note that, for planets with reasonable surface temperatures in orbit around hot, highly luminous stars, the brightest stars have the most distant life zones, so these two effects work in opposite directions to mitigate each other. As for explosive ejection, the violence of impacts depends on the relative velocity of impacting bodies, which is higher for more massive stars and lower at great distance from the star, where all orbital velocities are smaller. Again, planets of more massive stars, in order to have reasonable temperatures for the origin and development of water-based life, must be at greater distances, offsetting the effect of the star's mass.

As we have compared our Earthlet to Mars, we have in many cases profited from the insight provided by recent spacecraft exploration and observations of Mars. But these two planets are by no means identical. What sets Mars apart from Earthlet? Mars apparently formed with more volatiles, but with a much greater distance from the Sun and much lower global-average temperatures. Our Earthlet would, by definition, form with the same supply of volatiles as Earth, and would experience the same temperatures as Earth. Thus Earthlet would probably be significantly more vulnerable to loss of volatiles than Mars, because of both its closer proximity to the Sun (a higher intensity of ultraviolet radiation) and its smaller initial endowment of water and gases.

NOW CONSIDER a planet a little larger than Earth. Many of the factors operative in Earthlet's formation would, with respect to a larger body, generate a planet that is in some ways more Earthlike than Earth itself; that is, the planet would show in more extreme form many of the distinctive traits, such as massive oceans, that set Earth apart from the other planets. A larger planet with the same initial composition as Earth, let us call it Earthissimo, would have a greater average water depth than Earth, a higher acceleration of gravity, and consequently much higher pressures at the bottom of the oceans. The temperatures in its crust would increase more rapidly with depth, causing the crust to lose strength at shallow depths. That, plus the stronger gravity, would make the mountains much lower and the topography generally much more subdued. The higher heat flow would make the planet more geologically active, which, with a thinner crust, implies numerous small active volcanoes. Release of volatiles from the interior of the planet should be at least as efficient as on Earth, and recycling should be even more vigorous, suggesting a smaller mass of volatiles tied up in crustal sediments at any time. The higher escape velocity would further discourage atmospheric escape, perhaps even for helium. Both explosive blowoff by impacts and hot-atom ejection would be completely negligible.

For the sake of concreteness, let us imagine an Earthissimo with twice the radius of Earth. This would be a true behemoth of terrestrial planets, comparable in size to the heavy-element cores of the Jovian planets. It may, in fact, be too big to remain a terrestrial planet: if it had accreted to a size much larger than Earth while the Solar Nebula was still present, the gravity of such a body would have captured gases wholesale from the nebula, making the planet into a Jupiter with an extremely hot, dense, massive envelope of hydrogen and helium. But if the accretion process took much longer than the lifetime of the gaseous nebula, it would accrete in the absence of a supply of gases to form a giant terrestrial planet. This is what we shall assume for the moment.

What would our planet be like if it were twice as big? The planet would have much higher internal pressures, and therefore be denser than Earth. Earthissimo, with twice Earth's radius and diameter, would have eight times Earth's volume and about twelve times its mass. The acceleration of gravity would be 30 meters per second per second (three Earth gravities). Its escape velocity would be 27.4 kilometers per second, 2.45 times as high as Earth's.

The same concentration of water as Earth means twelve times as much water on Earthissimo. That water is spread over a surface area four times as large as on Earth, for an average depth three times as great as on Earth — about nine kilometers instead of three. But the high gravity and weak, hot crust imply much lower mountains: the highest should be about one-third the height of the highest on Earth. Those on Earth reach up to twelve kilometers above the abyssal plains, or nine kilometers above sea level: those on Earthissimo should reach about four kilometers above the abyssal plains. Picture it: peaks rising four kilometers from the ocean floor — in nine kilometers of water. A mountain climber standing on the highest mountaintop would have five kilometers (three miles) of water above his head! There is no reasonable prospect of any mountain peak emerging into the light of day. This is a true water world.

The pressure at the tops of the mountains would be great because of the enormous depth of the overlying ocean and the high gravity: about seventeen hundred atmospheres, compared with only three hundred on Earth's abyssal plains. On Earthissimo's ocean floor, the pressure is a staggering three thousand atmospheres.

Strange things happen to water at high pressures. It is a peculiarity of ice (only the rare metal bismuth behaves the same way) that this substance floats in its own liquid. Every other pure solid substance sinks in its own melt. High pressures compress ice and make it more closely resemble water. Ice-skaters wear skates with narrow metal blades that concentrate the skater's weight on a tiny surface area, producing pressures approaching one thousand atmospheres: at these high pressures, ice actually melts. The skater glides on a thin film of water that instantly freezes as the skate slides away and the pressure is relieved. Experienced ice-skaters know that, in very cold weather, skates sometimes don't work, because the compressed very cold ice is still below its melting temperature. The greatest depression of the freezing temperature (−22°C; −8°F) occurs for a pressure of almost exactly two thousand atmospheres. Beyond that point, pressures are so high that high-density forms of ice are stable. These forms, called Ice III, Ice V, and so on, are actually more dense than liquid water: squeezing them harder makes them *less* like water. They cannot be melted by increasing pressure. Therefore the freezing temperature rises steadily at pressures beyond two thousand atmospheres. At about six thousand atmospheres the freezing point is back up to 0°C (32°F). The melting point continues to rise with higher pressures, reaching the normal boiling point of water (100°C; 212°F) at about twenty-four thousand atmospheres. Water at this temperature and pressure would freeze, not boil!

On Earthissimo's ocean floor, temperatures below about −18°C would permit dense Ice III to form and settle on the bottom. If the surface temperature of the planet were, say, −19°C, a layer of normal Ice I about five kilometers thick would coat the planet, floating on a layer of dense, cold liquid water about three kilometers thick, in turn resting on about a kilometer of dense Ice III, which sits on the ocean floor.

But if Earthissimo were as warm as Earth, no Ice III would form. The reason is that water has yet another peculiarity. Although most substances uniformly expand as they are warmed and shrink as they get cooler, water actually reaches a maximum density when cooled to about 4°C. Below this temperature, further cooling causes the water to begin to anticipate the bloated crystal structure of ice: the density again decreases. Attempts to cool down the surface of a body of pure water cools the entire lake to 4°C before the surface can be cooled to any lower temperature. Further cooling from above then develops a layer of normal Ice I on the surface. Heating from above has essentially no effect on the temperature of deep waters: warmed surface water is buoyant and floats on top, resisting mixing with the cold, dense water beneath. The ocean depths on Earth are close to 4°C. Both the presence of abundant dissolved salts and high pressure can diminish and even wipe out this phenomenon: the salinity and depth of the ocean both matter.

The thickness of the crust of Earthissimo is determined by the amount of crust-forming material present in the planet. The crust, like water, has three times as much mass per unit area, averaging a thickness of ninety kilometers instead of thirty kilometers as on Earth. The thickness of the lithosphere, on the other hand, is determined by the temperature profile in the crust. The higher interior temperatures in Earthissimo define a temperature gradient that is about two and a half times as large as on Earth: melting temperatures are reached far closer to the surface than on Earth. This argument suggests a lithosphere that is only about thirty kilometers thick, instead of about seventy-five kilometers as on Earth itself. But this means melting must reach well up into the crust. Thus the ninety-kilometer crust is about one-third solid and about two-thirds partly melted. More of the crust is in the fluid asthenosphere than in the brittle lithosphere. On Earth, the lithosphere includes the entire crust (averaging 30 km) and a good slice off the top of the upper mantle (averaging another 45 km): the lithosphere floats on the upper mantle. The differences are profound. Many of these differences are subtle consequences of the pressure-dependence of the partial melting and fractionation behavior of rocks. But for our purposes, the most profound effect is that the crust is everywhere floating on a thick layer of soft, low-density, crustal-composition magma: dense basalt magmas of deeper origin will simply not be able to rise through the less-dense asthenosphere.

The very thin, warm crust is quite weak and very susceptible to drift. On Earth, much of the impetus for crustal movements lies in the fact that a slab of dense, basaltic ocean crust, as it sinks beneath the continental margins, boils off the volatiles from its coating of sediment. As the slab experiences growing temperatures and pressures, it continues to sink and in effect pulls more oceanic crust along after itself. But on Earthissimo, all the crust and all the upper asthenosphere are made of the same materials. Sinking of cold crust will be much less vigorous because the densities of solid and liquid are so similar, but the crust is so thin that sinking can occur more or less anywhere or at any time. Recycling of crust must be very efficient. The crust of the whole planet should be mobile, well-mixed, and young, with modest composition differences from one place to another.

Rising and sinking regions in the crust may closely reflect rising and sinking currents in the mantle. Large numbers of active hot spots and ocean-floor volcanoes are possible, and even probable. And there is but one "continent," a thin, easily cracked, and mobile layer which nowhere emerges above sea level. That continent may well contain hundreds to thousands of ever-shifting little platelets, adrift on the currents of the asthenosphere.

The atmosphere of Earthissimo is relatively immune to violent disturbances. Circulation of the global ocean will transport heat toward the poles, greatly diminishing temperature contrasts on the surface. Polar pack ice will not accumulate on continents or become grounded and immobilized in shallow seas, but will instead be conveyed by ocean currents back to low latitudes to melt. Reactive volcanic gases such as carbon dioxide, hydrochloric acid, and sulfur oxides would be extruded into the ocean under enormous pressures, where they will readily dissolve. The powerful atmospheric effects of large volcanic eruptions on Earth, such as enormous dust injections and formation of high-altitude sulfate aerosols, would be absent on Earthissimo. Also, large continental impacts on Earth can cause climate catastrophes on Earth by raising dense global dust clouds. But such events will occur on Earthissimo only for those very rare impacts that are violent enough to blast all the way through the ocean and excavate the ocean floor.

As mentioned earlier, escape of atmospheric gases is likely to be of very minor importance on Earthissimo because of its high escape velocity. Even solar-wind sweeping of light gases from the polar regions at high altitudes may be frustrated, because Earthissimo probably would have a higher magnetic field strength than Earth, enough to afford better protection against the depredations of the solar wind.

That Earthlet, Earth, and Earthissimo, all beginning with identical compositions, should turn into such different planets is a startling illustration of the principle that "more of the same is not the same." We have identified many size-dependent phenomena, of which the most striking trend is from a very few giant volcanoes on a drift-free, one-plate planet with isolated seas to an Earth with dozens of plates, several emergent continents in a global ocean, active drift, and multiple hot spots, to an Earthissimo with hundreds of jostling minor continental fragments, hundreds of moderate-sized hot spots, no basaltic ocean basins, and no emergent continents in its global ocean.

Weathering processes on the three planets also differ in fundamentally important ways. Earthlet, incapable of effectively recycling its thick, cool crust, irreversibly accumulates sediments in enormously thick layers. Earth sweeps weathering products into the seas, where river sediments and chemical and biological precipitates from the ocean fall onto swiftly moving oceanic crust, and volatiles are eventually (typically within one hundred million years) subducted and recycled as volcanic gases. Earthissimo churns its thin, hot ocean crust so rapidly that sediments do not accumulate to great depths, and all volatiles are expelled — into the ocean, not the atmosphere. Normal weathering by atmospheric oxygen and rainwater is replaced, on Earthissimo, by hydrothermal (hot-water) reactions: ocean water infiltrating deep cracks in the hot, rapidly changing ocean floor, and reacting there with minerals. The most rapidly weathering kinds of rocks found on Earth, basalts and other dense, iron-rich and silica-poor rock types, would be very rare or absent in Earthissimo's lithosphere. Mechanical weathering by wind and wave, running water, freeze-thaw cycles, and ice sheet movements, all of which are enormously important on Earth, would be completely absent on Earthissimo. Possibly, weathering is less important on Earthissimo, but we really cannot be certain.

The strength of the greenhouse effect on these three planets also varies greatly with size. Loss of atmosphere from Earthlet is the most important single factor. It remains unclear whether carbon dioxide and other volcanic gases would be very abundant on Earthissimo, because we have so little precedent for understanding the idea of removal of volatiles solely by hydrothermal action; therefore we cannot rule out an enhanced greenhouse effect and elevated atmospheric and oceanic temperatures. Elevated temperatures cook dissolved volcanic carbon dioxide out of the ocean and enhance the greenhouse effect, which further warms the ocean — leading, quite possibly, to a runaway greenhouse effect and Venuslike surface temperatures. Whether this happens probably depends very sensitively on the planet's distance from the Sun. Here, as in so many other cases, our own Solar System is woefully short of useful examples. Planetary scientists in fact disagree on whether this is what actually happened on Venus.

Escape is a serious issue for Earthlet, and much less so for its larger sisters. Impacts on Earthlet strip away atmosphere; on Earth and Earthissimo, they normally result in a net addition of volatiles. Large impacts should be exceptionally common and violent on Earthissimo, but the deep protective blanket of ocean and the highly mobile crust should not only inhibit cratering, but also quickly destroy any craters that do form. Large impacts on Earthissimo would generate massive, broad tsunami waves that would course uninterrupted many times around the planet, never shoaling enough to build up to great heights and break. The major ecological effects of impacts would involve raising steam to high enough altitudes to destroy the ozone layer, or shocking the atmospheric nitrogen-oxygen-water-vapor mixture to make abundant nitrogen oxides and acids.

And what about life? From our experience on Earth we have no reason to rule out an early origin and evolutionary development of life on Earthlet, but any life surviving for billions of years would have to be adapted to the few congenial environmental niches. Life on Earth we are intellectually prepared to accept (even if there is much we do not understand). Life on Earthissimo may well be a phenomenon centered on ocean-floor hydrothermal vents, where hot, nutrient-laden streams of water, cooked out of ocean-floor sediments, gush out into the relatively cold ocean. As at hydrothermal vents on Earth's ocean floor, the course of life there would differ in many distinctive ways from that on a planetary surface. Because of the likely transient nature of these vents and their probable multiplicity, reproduction of surface-bound organisms (the local equivalents of clams, worms, crustaceans, and plants) would probably involve an early free-swimming stage to permit escape when the local vent dies.

Obviously, there are many possible Earths that lie between the extreme cases we have developed here. A planet with two or three times the mass of Earth might have numerous volcanic islands emerging from a global ocean. A planet of a third of Earth's mass, intermediate between Earthlet and Earth, might be half-covered with water, bear far more visible impact scars than Earth, and retain enough atmosphere for life to persist for billions of years while continuing to develop to more complex, more capable forms.

Now is a good time to remind ourselves that size is not the only factor that distinguishes the fate of grossly Earthlike planets. It is also useful to consider planets that are "slight" variations on the composition of Earth. Two examples of such closely related, but compositionally distinct, planets are Venus and Mars.

VENUS, AS we see it today, differs most strikingly from Earth in that it is an extremely dry planet, a red-hot global desert. Since Galileo asserted the possibility of water on other worlds, and Campanella defended these ideas with citations from science and Scripture, few more depressing sights than Venus have come to light: a grossly Earthlike planet with virtually no water! How did this come about? Terrestrial theorists differ widely in their reconstructions of the evolutionary history of the planet. Although there is no proof that Venus started out with the same composition as Earth, one school of thought, currently led by Thomas Donahue of the University of Michigan and Yuk Yung of the California Institute of Technology, holds that Venus started with about the same amount of water as Earth and lost that water through photolysis and massive escape of hydrogen. Although there is no proof that Venus started out with a composition different from Earth, an opposing school of thought, currently led by David Grinspoon of the University of Colorado, and myself, holds that the amount of primordial water on Venus cannot be deduced from the present state of the atmosphere, since the water abundance and hydrogen isotopic composition can plausibly be explained by late infall of cometary and asteroidal water onto a Venus that may have started with almost any content of water, even none at all. Until we know which of these two theories is closer to the mark, drawing reliable conclusions about the primordial composition and evolutionary history of Venus will remain impossible.

Mars today is an ice planet: it is so cold that liquid water, even as a brine, is extremely rare or absent on its surface. Yet powerful geological evidence reveals a warmer, wetter past. Several lines of evidence also suggest a higher atmospheric pressure in the distant past. But both the total amount of volatiles on Mars today and the amount present billions of years ago are uncertain. Theories of the origin of the Solar System tend to assign a higher water content to Mars than to Earth, but we cannot prove that this is true because whatever water is present there today is frozen and buried. Also, the ease with which atmospheric gases are lost into space suggests that escape is very important, which makes reconstructing the original composition a very uncertain art.

For both Venus and Mars, evolutionary changes are of great importance. Loss of ancient water is possible on Venus due to a powerful greenhouse effect and rapid UV photolysis, as is late addition of water by the infall of volatile-rich comets and asteroids. Explosive blowoff and dissociative recombination encourage loss of atmospheric gases from Mars. These effects are important on Venus because of its proximity to the Sun, and on Mars because of the low planetary mass. Size matters; composition matters; location matters.

Is Earth, then, the best of all possible worlds? Is it the best of all possible Earths? For whom? How Earthlike must a planet be before we consider it as a possible abode of Earthlike life? What about life that is not terribly Earthlike? Do we in fact really need highly Earthlike planets in order to have life prosper and evolve to complexity? At this point we can only dare a guess; but I will take the dare and express my personal expectations. I think Earth is a wonderful example of an inhabitable world; but I also think the range of congenial worlds is far wider than the range of those that humans would desire to live on. A planet does not need to be terribly Earthlike to harbor life, especially when one allows for life forms that evolved to harmonize with their native environment, rather than being arbitrarily (and unwisely) assumed to be identical with forms that arose on Earth. Planets, to be congenial to life, really should have lots of liquid water, and they really should not be too hot, too short-lived, or too unstable in their environmental conditions. Earth fits the bill—but so do lots of other very un-Earthly places. And Earth life, like all other forms of life in a living Universe, knows how to evolve. Transport an ecosystem from one planet to a rather similar one, and that ecosystem will adapt over time, through natural selection, genetic variation, and mutation, to better fit the local environment.

One clear result of this exercise is an increased tolerance of planets that look a lot different from Earth. Even the complete absence of dry land need not preclude an interesting and diverse biosphere (although the complete absence of water is a serious problem). We are certainly encouraged to look at a broader range of conditions, and a broader range of types of planets, than simply those that mimic Earth. We should consider giant planets, large satellites, and even brown dwarfs, in addition to grossly Earthlike planets: we must seriously contemplate bodies that begin with compositions very different from Earth. To do so, we should recognize and reject "terrestrial chauvinism"—if it is in fact possible to do so. Such attitudes are subtle and pervasive: we constantly find ourselves diverted into discussions of how "Earthlike" a planet is, as if that were the only consideration that matters. We need to see through this semantic screen and establish our own standards and purposes: we seek not just other Earths, but other worlds capable of supporting their own native life. Our ability to do so is inevitably impaired by traditional wisdom and the conventions of language, but we can to some degree liberate ourselves from these constraints by articulating and questioning our tacit assumptions. Even so, we must expect the richness and variety of reality to surpass our wildest flights of inventiveness.

9

A FAMILY OF GIANTS

OF THE DOZEN-PLUS GAS-giant planets known, only a minority are in our own Solar System. Those four, Jupiter, Saturn, Uranus, and Neptune, have all been studied intensively by astronomers and have all been visited by spacecraft from Earth. These few relatively well-understood examples appear to fall into two natural groups, distinguished by mass, distance from the Sun, and composition: Jupiter and Saturn have masses of about one hundred Earths or more and are rather close in composition to the Sun; Uranus and Neptune, with masses of a dozen Earths or so, are made of intrinsically denser material. There is a significant gap in both mass and composition between these two groups. True, it is possible that we are simply looking at four more or less random samples from a continuum of giant planet types, and observation may soon fill in the gap between them. But for the moment it is convenient to treat them as two groups, Jovian (Jupiter and Saturn) and Uranian (Uranus and Neptune). We shall discuss these two classes in sequence, for the purpose of understanding the environments that they provide for the origin and development of life.

Jovian planets are made from small solid planetesimals of dirty ice, in which all the rock- and ice-forming elements are present in solar proportions, mixed into a much larger mass of gaseous elements (principally hydrogen and helium) gravitationally captured from the Solar Nebula. If all the available gas were captured, the final planet would have exactly the composition of the early Sun. Since the capture process is not perfectly efficient, the planets made in this way are depleted severalfold in gaseous elements, relative to the composition of the Sun. Alternatively, we could say that all the ice- and rock-forming elements are enriched severalfold relative to hydrogen and helium. Jupiter, for example, has roughly five times as high a concentration of "heavy" elements (carbon, nitrogen, oxygen) as the Sun. Saturn is enriched in these heavy elements by about a factor of ten. The heavy-element ("mud") cores of both planets have about the same mass, even though Jupiter has a total mass (318 Earths) that is more than three times that of Saturn (around 95 Earths).

Because of their high interior temperatures, Jovian planets are generally fluid (gas or liquid) throughout. As heat is radiated away from their topmost cloud layers into space, the energy content of the planet declines, lowering the temperature. Lower temperatures cause a decrease of the internal pressure, which allows the entire planet to shrink slightly under the influence of its own gravity. The smaller planet has a higher acceleration of gravity at every altitude. The gravitational potential energy change associated with shrinkage compresses the gas and heats it, which keeps convection going throughout the outer atmosphere, carrying internal heat up to the cloudtops, where the heat can be radiated away into space. In effect, a Jovian planet several billion years old is still playing out the tail end of the formation-and-collapse process that began in the Solar Nebula.

Collapse actually causes the temperatures and pressures deep within the planet to increase. In fact, the center of Jupiter must be far hotter than the visible face of the Sun. Jovian planets can typically maintain temperatures of 10,000 or 20,000 K in their cores, hot enough to volatilize anything, but still very far short of the temperatures in the deep interiors of stars (usually 10,000,000 to 20,000,000 K) where nuclear reactions take place. Fusion reactions are therefore incapable of providing any significant internal heat in a Jovian planet. To maintain high enough central temperatures for even the most tentative fusion reactions to occur (fusion of the heavy hydrogen isotope deuterium), a planet would have to have a mass of at least 13 Jupiters (more than four thousand Earths).

Because Jovian planets have an ongoing internal source of heat, they radiate off more heat than they absorb from the Sun. On Jupiter, for example, the Sun provides only 12.7 watts of sunlight per square meter, averaged over all latitudes and over the day-night cycle. About 30 percent of this light is reflected off Jupiter's clouds and lost back into space, leaving 9.2 watts per square meter to be absorbed and turned into heat. But the observed amount of heat coming from Jupiter's cloudtops is 14.7 watts per square meter. Of this, the Sun contributes 9.2 and the escaping internal heat of Jupiter contributes 5.5. On Saturn the situation is similar, but, because Saturn is farther from the Sun and less massive than Jupiter, all the numbers are smaller: 3.8 watts per square meter of sunlight, of which 2.9 is absorbed and converted to heat, with a total observed heat output of 4.9, of which 2.0 is escaping internal heat. For comparison, the intensity of sunlight at the top of Earth's atmosphere is 1,370 watts per square meter, and the average input over the entire planet is 343. Earth's surface is maintained at an average temperature of about 300 K ("room temperature"). Earth's internal heat source is so feeble that the planetwide input and output balance almost perfectly.

The much lower heat emissions of the Jovian planets suggest very much lower temperatures. In fact, Jupiter radiates like a perfectly black sphere with a temperature of 124 K (on the Kelvin temperature scale, which counts from absolute zero using Celsius degrees). In somewhat silly and impractical units, that is a very chilly −236°F, far below the lowest temperatures ever measured on Earth. Saturn is an even colder 95 K (−288°F).

So low are these temperatures that almost everything in their atmospheres condenses to form solid cloud particles. The topmost clouds of Jupiter are made of ammonia snow. The even colder polar regions of Jupiter have a tenuous high-altitude haze of methane and ethane snow. Above the main cloud tops of Jupiter the only gases remaining uncondensed in the atmosphere are hydrogen, helium, neon, and methane, plus tiny traces of several partly-condensed gases such as ammonia. Saturn is similar, but everywhere colder.

The temperature in any Jovian planet increases with depth in its atmosphere. Not too far beneath the cloud tops, mild temperatures and clouds of water droplets can be found on both Jupiter and Saturn. The water-cloud droplets, immersed as they are in an atmosphere containing highly soluble ammonia gas, are actually a dilute solution of ammonia in water. This basic (ammonia-bearing) solution in turn readily dissolves hydrogen sulfide, a minor component of hydrogen-rich atmospheres. The water-cloud layer is for the most part surmounted by a layer of ammonium hydrosulfide snow, made by reaction of cold ammonia and hydrogen sulfide gases, and that cloud layer is in turn mostly covered by the topmost layer of ammonia crystal clouds. Observers looking at Jupiter from the outside generally see the high, white ammonia clouds covering about 90 percent of the planet. The brownish ammonium sulfide clouds cover almost all of the rest, but local breaks in that cloud layer often permit us to see small areas that are clear all the way down to water clouds at room temperature, where the atmospheric pressure is around six atmospheres. Liquid water clouds, with temperatures suitable for life as we know it on Earth, provide Jupiter with better than 99 percent global cloud cover. Since the surface area of Jupiter is some 120 times as great as the entire surface area of Earth, this means that the sum of all clear, cloudless areas on Jupiter is comparable in extent to Earth's surface. It was in one of these rare cloudless areas that the 1995 Galileo entry probe chanced to fall.

Below the base of the water clouds, the temperatures continue to climb, evaporating water droplets and clearing away the clouds. A hundred kilometers below the water clouds the temperatures will be high enough to destroy delicate organic matter and sterilize the atmosphere. Much deeper in, at much higher temperatures, we would begin to encounter clouds of condensed rocky materials, vapors rising from the even hotter interior.

The internal heat that we see emerging from the top of the atmosphere cannot get there as radiation: the dense lower atmosphere is too opaque. Instead, the heat is carried by rising convection currents, like air over a warm stove. Over most of the planet, the air is warm and rising. But the rising air must be balanced by an equal mass of subsiding cool air. The counterflow of rising and sinking currents necessarily generates turbulence, which mixes cold, dry air from high altitudes with hot, moist, sterile air from the Stygian depths.

Sunlight scarcely penetrates to the water-cloud layer. Not only do the overlying cloud layers obscure the Sun, but even the cloud-free atmosphere above clearings in the lower clouds can hinder the passage of sunlight. The same Rayleigh scattering phenomenon that makes Earth's sky blue and the setting Sun red works with a vengeance on Jupiter. With an atmospheric pressure of six atmospheres at the water-cloud layer, the Sun would always appear reddened. Much of the blue and violet sunlight never reaches the clouds: it is scattered so efficiently, and redirected by scattering so many times, that much of it escapes back out to space. Indeed, when we observe Jupiter from Earth or from a passing spacecraft, the regions where there are holes in the upper clouds and where water clouds are exposed to view are always very blue. This does not mean the water clouds are themselves actually blue; only that there is so much atmosphere above them that the blue part of the sunlight gets scattered back toward the Sun without ever reaching the water clouds.

Solar ultraviolet light, with an even shorter wavelength than blue light, is much more efficiently scattered. Little of it gets past the ammonium hydrosulfide clouds, so that there is essentially no chemically active ultraviolet light available at the level of the water clouds. But scattering is not the only threat to passage of ultraviolet light: it is also strongly absorbed by certain important molecules. On Jupiter, several gases that are sensitive to attack by ultraviolet light are actually present high enough in the atmosphere to encounter UV sunlight, absorb it, and thus shield the lower atmosphere from its effects. These gases are methane, ammonia, phosphine, and hydrogen sulfide. Water vapor condenses so deep in the atmosphere that it never encounters ultraviolet light capable of dissociating it. This process of tearing apart a molecule by means of high-energy light is called photolysis, from the Greek words meaning "dissolving by light." Photolysis provides a variety of reactive molecular fragments that readily combine to produce more complex, even biologically interesting, products.

Basically, the longest-wave ultraviolet radiation is the least energetic, the least effective at photolysis, the most deeply penetrating, and by far the most abundant. In the vicinity of the ammonium hydrosulfide clouds, long-wavelength ultraviolet light, well beyond visible violet light in the spectrum, can just reach hydrogen sulfide gas. Hydrogen sulfide has unusually weak, fragile chemical bonds. The encounter is destructive to both parties: when the ultraviolet photon is absorbed by hydrogen sulfide, its energy wrenches off a hydrogen atom. This event initiates a cascade of reactions that produce sulfur and polysulfide chains that are yellow, orange, red, or brown in color. The ammonium hydrosulfide clouds themselves, initially colorless, are also vulnerable to even longer-wave UV light, just short of the limit of human vision. More of the same suite of color-bearing molecules (called "chromophores") are produced, staining that cloud layer an unattractive orange-brown. When we look at this cloud layer through breaks in Jupiter's ammonia cloud layer, we in fact see just such colors.

Slightly shorter-wave (more energetic, and more easily scattered) ultraviolet radiation is capable of tearing a hydrogen atom off an ammonia molecule. This usually occurs at the highest altitude at which ammonia gas is abundant, which is in the tops of the ammonia cloud layer. The main product of ammonia photolysis is nitrogen gas, but some of the NH_2 fragments made by the breakup of ammonia can combine with each other to make hydrazine, a very poisonous liquid with physical properties similar to water. Some of the hot H atoms kicked out by photolysis can also strip hydrogen atoms from methane, making a little methylamine. Although technically an organic compound, methylamine is of negligible biochemical interest.

An important side product is hydrogen cyanide (HCN), which is a precursor of a variety of organic molecules that are essential to all known life forms. All that hydrogen cyanide needs to make it into amino acids, organic bases, and other crucial biochemicals is water. In other words, as HCN and other water-soluble and reactive organic materials settle down through the atmosphere to the vicinity of the liquid water clouds, they may dissolve in rain droplets and pursue further chemistry.

Phosphine, very similar in structure and bond strength to ammonia, is present in trace amounts in the same part of the Jovian atmosphere where ammonia is dissociated. Phosphine, when torn apart by UV light, makes a rank-smelling flammable and neurotoxic gas called diphosphine and a family of multicolored (red, yellow, and black) solid forms of phosphorus. These can be potent coloring agents for the clouds of Jupiter, but the higher atmospheric density on Saturn may suppress both phosphine and ammonia chemistry by scattering away the incoming UV light. There is no obvious biochemical significance to these reactions and products.

At even higher altitudes, even more energetic (and hundreds of times less abundant in the Sun's spectrum) ultraviolet light tears apart methane molecules to make a variety of simple hydrocarbons, such as ethane, ethylene, acetylene, propane, and so on. Of these, acetylene is most reactive and most capable of further reactions that make biologically interesting materials.

Thus there are potential precursors of biologically interesting material that are made at high altitudes on Jupiter and Saturn by known reactions. Vertical mixing of the atmosphere, which must take place because of the turbulence driven by internal heat sources, can transport tiny concentrations of these products down to the water-cloud layer, where further reactions can occur. There is a feeble but credible source of complex organic matter on the Jovian planets, sufficient to explain the traces of HCN and hydrocarbons detected by spectroscopists and by the Galileo entry probe. But only some 0.05 percent of the photolysis reactions on Jupiter involve a carbon-bearing (methane) source and organic products. The rest are inorganic products of hydrogen sulfide, ammonia, and phosphine.

URANUS AND Neptune differ from Jupiter and Saturn in several ways. First, they are farther from the Sun: the supply of solar radiation reaching them is diluted by their immense distance from the central fires of the Solar System. As a direct result, they are colder than Jupiter and Saturn at all pressure levels. Second, Uranus, for rather arcane reasons, has no detectable internal heat source, although Neptune does. Third, the Uranian duo contain about 50 percent of icy and rocky material, compared with just a few percent for the Jovian planets: the heavy elements are enriched in them by a factor of roughly one hundred. Interestingly, despite the wide diversity of the masses of the giant planets, the heavy-element cores of all four are closely comparable in size.

Imagine adding ten times as much of the condensable elements to Saturn and cooling the whole planet down. All the cloud layers would form deeper in the atmosphere. Water clouds, forming from an atmosphere with a much higher density and a ten-times-higher water concentration than on Saturn, would be deep and dense. The water-cloud base would consist of large, hot, precipitating droplets of ammonia solution. The topmost cloud layer from Saturn would now form at pressures of several atmospheres, instead of about 0.7 atmospheres as on Jupiter, and the temperature would continue to drop off above that level. By the same method we used on Jupiter and Saturn, we can measure the heat emission from Uranus and Neptune and see just how cold their cloud tops are. The answer is that both planets are at almost exactly 59 K (−353°F). If Earth were this cold, both nitrogen and oxygen would condense and fall as snow. On Uranus and Neptune, where nitrogen and oxygen gases are absent, methane condenses and forms a new topmost cloud layer, masking the ammonia clouds.

Ultraviolet radiation from the Sun has great difficulty penetrating either the methane clouds or the very dense atmosphere. Absorption and scattering would protect ammonia, phosphine, and hydrogen sulfide from attack. Methane would be vulnerable to attack, but only at a rate limited by the much lower intensity of sunlight on the Uranian planets, caused by their much greater distance from the Sun. For example, the flux of UV light onto Uranus is only 6 percent of that on Jupiter. Because only methane is exposed to photolysis, only hydrocarbons will be made by ultraviolet sunlight striking the atmospheres of Uranus and Neptune, and they will be made at extremely low rates. Our discussion of Jupiter and Saturn suggests that some of these organic products might react in the deep, dense and hot water clouds to make interesting and useful organic materials; however, hydrocarbons are relatively unreactive and insoluble in water. Uranus and Neptune are therefore far less congenial to the production of biologically interesting material than are Jupiter and Saturn.

BY TOURING the Jovian and Uranian sets of twins, we have seen that three important factors enter into determining the planetary environment and its suitability for the origin of organic material and the rise of life. These three factors are the composition of the planet (essentially, the percentage of heavy elements), the distance of the planet from the Sun (in more general terms, the intensity of sunlight striking the planet), and the mass of the planet. These are in fact the exact same factors we earlier identified as the crucial determinants of the properties of rocky worlds.

These three basic variables are not necessarily completely independent of each other: very gas-poor compositions seem to be a natural result of accretion of the planet very far from the Sun, where the density of the nebula is low and long times are thus required for the assembly of solid, gravitating planetary cores. In the absence of massive gas capture, these planets should be low in mass as well as cool and distant.

This tour also prepares us to understand giant planets in an even broader variety of settings, including orbits around other stars. But when we consider other stars, a fourth major factor enters in: the temperature of the central star. Its temperature determines the spectrum of starlight falling on its planets, especially the abundance of UV radiation. Hot stars have proportionately more UV; indeed, the most luminous main sequence stars, those of O and B classes, can emit most of their energy in the ultraviolet. Fainter, cooler main sequence stars, of spectral classes K and M, may emit far less than 1 percent of their energy as UV light.

The range of composition possible for giant planets seems defined by the idea that they are mixtures of two distinct components, one a solid ice-rock mixture of condensed materials, resulting in what is sometimes referred to as a "dirty snowball," and the other a complementary mixture of "permanent gases," almost entirely consisting of hydrogen and helium. The solids, once accreted into Mars-size bodies, can capture some fraction of their complement of gases, producing planets with a range of composition that extends from pure "mud" to solar composition. The case of planets with exactly solar composition differs so slightly from Jupiter and Saturn that it does not need to be discussed separately; however, the nearly gas-free case is quite a bit more extreme than Uranus and Neptune. We could imagine planets with 10 percent, 1 percent, or less hydrogen and helium, effectively ocean planets with moderately high atmospheric pressures (thousands, not millions, of atmospheres) and enormous supplies of water, methane, ammonia, other volatiles, and rock. Deep oceans of water filled with dissolved salts, perhaps several thousand kilometers deep, would exist at such high pressures that they would start to freeze from the bottom up into high-density forms of ice. A thick ice layer would be a poor conductor of heat (especially compared with a freely convecting liquid ocean), so a large temperature difference would build up across the insulating dense-ice layer, leading to partial melting and thinning of the layer. This may well be a stable self-regulating process. Note, however, that this modified giant planet is beginning to look a lot like Earthissimo.

The division of raw solar-composition material into condensed solids and permanent gases of course works differently at different nebular temperatures. In the warm inner parts of the nebula, where the icy components are partially or fully evaporated, the two components would be rocky solids and gases, in which the gas mixture would include not only hydrogen and helium, but also water and many compounds of nitrogen and oxygen. But such high temperatures both decrease the mass of condensed solid (by vaporizing it) and increase the amount of solid mass that must be accumulated to bring about gravitational gas capture. Therefore it seems unlikely that this kind of scenario can be pushed to temperatures above the ice condensation point.

Colder nebular temperatures, on the other hand, would have little or no effect: at temperatures below about 30 K, the hydrogen-helium mixture contains only one other gaseous ingredient of any significance, neon. Temperatures low enough to condense neon or hydrogen seem impossible to attain in the vicinity of a young star, and helium condensation requires temperatures so low that they are below the background temperature of the Universe maintained by the leftover radiation from the Big Bang, and therefore unachievable.

So far we have considered only solar-composition starting material. But, for the Galaxy at large, not all stars have the same composition. Their planets therefore may also differ from the giant planets in our own system. It is true that most main sequence stars now present in the spiral arms of our galaxy, the so-called Population I stars, have compositions very similar to the Sun. But it is also true that the composition of raw stellar material changes with time. At the time of the origin of the Milky Way, the raw material available was exclusively debris from the Big Bang explosion, essentially a mixture of hydrogen and helium with no other significant ingredients. The earliest stars must have formed with virtually zero content of all elements heavier than helium (what astrophysicists glibly mislabel "metals"). Some of these first-generation stars must have been massive enough to stand briefly at the upper end of the main sequence. They burned hydrogen at a furious rate, then blew themselves apart in supernova explosions while other first-generation stars around them were still slowly forming. Continuing star formation causes continuing supernova explosions, each enriching the interstellar gas with large quantities of heavy elements. Newly forming stars sample this mixture of Big Bang gas and fresh supernova debris, so each new generation of stars that forms begins its life richer in heavy elements. Star formation in the spiral arms of the Galaxy began early and continues today; indeed, stars now forming draw from an interstellar gas that is richer in heavy elements than the raw material that was available for the formation of the Solar System 4.6 billion years ago.

Interestingly, there are parts of our galaxy in which there are no gas and dust clouds, and in which there has been no star formation for at least ten billion years. These are the gas-free globular clusters, spheroidal clusters of thousands to millions of stars that orbit in randomly oriented, elongated elliptical orbits around the galactic core. Twice on each orbit these clusters cross the galactic plane, where they are stripped of gases and dust by passing at high speeds through the dense spiral arms. The stars in globular clusters have populations that differ strikingly from the familiar Population I spiral-arm stars that we see orbiting near us in the galactic plane. These globular cluster members, the Population II stars, have very low abundances of heavy elements. They also occupy only the lower half of the main sequence: star formation in them ceased so long ago that all the short-lived members have already evolved off the main sequence and become giants, supergiants, or supernovas. In fact, the globular clusters contain no stars with lifetimes less than about eight to ten billion years.

Some of the clusters turn off the main sequence at the point where stars have main sequence lifetimes as long as twelve to fifteen billion years. Such stars are exceptionally "metal-poor" in the language of the astrophysicists: they have abundances of carbon, nitrogen, oxygen, silicon, iron, and so on that are hundreds to thousands of times lower than in the Sun and its fellow Population I stars.

Consider the formation of a planetary system from the nebula out of which an early, metal-poor Population II star was forming, perhaps twelve billion years ago. The amount of ice-forming and rock-forming elements was about a thousand times lower than we had in our Solar Nebula: there is neither enough rock to make terrestrial planets, nor enough ices to trigger the formation of Jovian and Uranian planets by gravitational capture. Planets of stellar composition might be made by rotational splitting of forming stars that, as they shrink, come to spin too fast to be stable, but they cannot form by the mechanisms familiar to us in the Solar System. If gas-giant planets do not form by this rotational fission process, then no other route would be available: the metal-poor star's system would resemble an immense, sparse asteroid belt. The biggest bodies in it would probably be comparable in size to Ceres, about one thousand kilometers in diameter. Such bodies cannot develop or retain either atmospheres or liquid water.

Since star formation has been going on in the spiral arms of our galaxy for about twelve billion years, some of the nearby stars must be very ancient. The vast majority of them, however, are much younger, and very diverse in age. But consider: those spiral arm stars that are older than ten billion years (and are therefore moderately metal-poor) have aged to the point at which all stars more massive than the Sun, those with main sequence lifetimes of ten billion years or less, have already evolved off the main sequence. The surviving ancient spiral arm stars will be a very unassuming assortment of intrinsically faint stars. Some fraction of the G, K, and M stars we see are certainly much older than the Sun; we can tell which ones they are by the fact that they are metal-poor, which means that they must have formed a very long time ago. Conversely, recently formed stars of these same spectral classes may have metal contents as much as twice that in the Solar System. They could produce systems of planets that are significantly more massive or more numerous than in the Sun's family. Their Jovian planets could potentially be more massive and denser than ours, although generally chemically similar in all qualitative respects.

What about gas-giant planets that find themselves in orbits that force unusually high or low temperatures on them? Excursions to lower temperatures will produce bodies with all the chemical simplicity of Neptune or Uranus: atmospheres of hydrogen, helium, and neon, possibly containing argon and methane if they are as warm as fifty-five or sixty degrees. Cloud-top temperatures below fifty Kelvins would freeze methane and argon out of the atmosphere, leaving only hydrogen, helium, and neon. Since helium, neon, and argon are virtually impossible to detect by remote observations, such a planet would appear as a dark blue, cloud-free, featureless ball of hydrogen gas. As light from its sun penetrates the planet's atmosphere, the short-wave light would be wiped out by scattering, and long-wave (red and infrared) light would be absorbed by the cold, dense hydrogen gas.

Gas-giant planets much warmer than Jupiter are certainly conceivable, and might be quite interesting bodies. Suppose that the luminosity of the star and its distance were just right to set the temperature of the planet's surface at 180 K. The internal heat source of the planet would raise this modestly to about 200 K. Then ammonia clouds would not condense, except perhaps in the immediate vicinity of the poles. The topmost cloud layer would be ammonium hydrosulfide, which, exposed directly to almost unattenuated ultraviolet light, would be stained a deep orange-brown by photolysis. Fresh new clouds would be shades of yellow and orange. Red features made of phosphorus are unlikely, since they would probably be overwhelmed by sulfur. Large breaks in the ammonium hydrosulfide clouds would expose pale bluish cloud systems made of water and ice.

Warming the planet further to 230 or 250 K would effectively dispel the ammonium hydrosulfide clouds, leaving a dense water-cloud layer directly exposed to sunlight, constantly seeded by photochemical products of methane, ammonia, phosphine, and hydrogen sulfide. The water clouds would be a seething hotbed of chemical activity. Further warming to 320 or 350 K would completely dispel the water clouds, leaving the atmosphere devoid of opportunities for water-based chemistry. Even higher temperatures are conceivable, but the biological interest of such bodies is greatly diminished by the absence of liquid water, which is essential to familiar organic chemistry.

Finally, we may imagine the most extreme size variants of giant planets. Masses below about ten Earth masses are possible, containing little captured gas. These are variations on the global-ocean planets previously mentioned, akin to both Earthissimo and the Uranian planets. The question of how massive a gas-giant planet can be is actually more a question of semantics than physics. Most astronomers would define any body more massive than 0.07 Suns as a star. Bodies in the size range from 0.07 Suns down to 0.01 Suns are often called brown dwarfs. We shall deal with brown dwarfs more extensively in the next chapter. Bodies from 0.001 Suns (Jupiter) to 0.01 Suns (13 Jupiters) are simply large gas-giant planets. The latter dividing lines are, however, a matter of taste. For now, let us simply remark that bodies several times the mass of Jupiter have strong enough gravitational attractions to compress their gaseous interiors even further. Because of self-compression, the actual size (diameter) of the planet does not grow much as the mass increases from one Jupiter to 13 Jupiters. The acceleration of gravity, however, rises by about a factor of ten over this size range, making the atmosphere much more compact, with a steep rise of pressure (and temperature) inward from the cloud tops. The amount of internal heat generation also rises steeply over this size range. Since the surface area is grossly constant, this means that the "surface" (cloud-top) temperature of the planet becomes more and more dominated by the internal heat loss, and less and less sensitive to heating by the planet's sun. When the internal heat source becomes dominant, we have effectively a bogus star, a body at temperatures of hundreds of degrees, glowing in the infrared, drawing its power not from nuclear reactions but from its internal shrinkage. Beyond this point, the name *brown dwarf* seems appropriate.

Gas-giant planets of the right size and age may therefore maintain cloud-top conditions at temperatures in the liquid water range *even in the absence of a nearby star*. This is surely the height of planetary self-sufficiency! But even this halcyon state of affairs is only temporary: over the long term, the "surface" of the planet continues to cool slowly, driving the layer with benign temperatures ever deeper into the atmosphere, to eventually be deeply covered with dense, high-altitude clouds of ammonia and other materials.

Are there stable ecological niches in any of these giant planets? At any given time, there are certainly altitude ranges that have temperatures in the liquid water range on many of these bodies. But all of them are strongly affected by turbulent vertical mixing driven by the escape of internal heat. There are highly desirable places in the atmosphere, but nothing stands still. A parcel of atmosphere that is at this moment at room temperature and filled with fat, warm droplets of liquid water, dripping with photolysis products containing hundreds of compounds of carbon, hydrogen, nitrogen, oxygen, sulfur, and phosphorus, will in a few days be exposed to killing ultraviolet radiation at the frigid temperature of 100 K, or cooked at pressures of thousands of atmospheres and temperatures of thousands of degrees, completely destroying all interesting molecules. The problem is that the origin of life does require time: complex molecules must be built up stepwise, in vast numbers, and natural selection must choose the winners and losers from the competitors. But if rapidly changing conditions totally destroy (by heat) or immobilize (by cold) these chemicals and their reactions, there can be no winners. The origin and development of life on a giant planet is by no means a foregone conclusion. While it is certainly possible to speculate, as Carl Sagan did, that buoyant living "gas bags" might float in the atmosphere of a Jovian planet and regulate their altitude (and temperature) by adjusting their buoyancy, there is no credible evolutionary path to such creatures. I have long maintained that such life forms could be supported only by hot air.

IF GIANT planets exist in orbit around nearby stars, how would we ever know it? Fortunately, over the past few years it has become technically feasible to detect such planets. There are several possible techniques that can be brought to bear. Each technique has its own challenges, strengths, and blind spots, and each is selective in different ways.

The first technique is to measure the positions of relatively nearby stars against the background of much more distant stars with very high precision. This is called *astrometry*. After months or years of observation, it is sometimes found that a nearby star is not following the simplest possible path across the sky. The simplest case is that of a single star without any large planets, which follows an orbit different from that of the Sun. Its apparent path in the heavens, correcting for Earth's annual motion around the Sun, is a straight line. But several nearby stars follow more complex paths, with wavy components having periods of not only one year (due to Earth's motion around the Sun) but also other periods. Such observations have been made with adequate precision for only the last few years; therefore, planets that have orbital periods longer than a few years are very difficult to detect—they have not yet completed so much as a single orbit since they came under observation. Any wavy component in the motion of another star can be due only to the fact that another, massive, but much fainter bodies orbiting that star. The center of mass of that stellar system follows a straight line through space, but neither the star nor its unseen companion lies at the center of mass: they both orbit that center. Therefore the star appears to follow a wavy path through the sky. The cycle time of the wave pattern gives the orbital period of the dark companion, whereas the amplitude of the wave (how much it departs from straight-line motion) depends on the relative masses of the star and companion; a more massive companion means the star is farther from the center of mass of the system, and the amplitude of the wave is larger.

The second technique, and the most successful to date, is based on detection of cyclic radial-velocity variations of stars. This is called the *radial-velocity method.* Even though making sufficiently precise velocity measurements is extremely challenging, the basic idea is rather simple: since the orbital motion of a planet causes a periodic motion of its star around their common center of gravity, the line-of-sight velocity of the star varies cyclically by a very small amount. The radial velocity, which causes a Doppler shift in the wavelengths of every line in the spectrum of the star, is largest if the ratio of the mass of the dark companion to that of the star is large. It is also large if the orbital period of the companion is short (that is, if its orbital velocity is high). Together, these factors favor easiest detection of very close, very massive planets. The variations of the velocity over time can even give us a quite precise knowledge of the eccentricity of the planet's orbit. There is, however, a subtle blind spot in this technique. Consider a planet orbiting a star in such a way that Earth (and the Sun) are close to the pole of the orbit. From Earth, the astrometric technique would see the star moving in a circle or ellipse on the plane of the sky. But, because the motion of the planet and star lie in a plane that is at right angles to our line of sight, the radial (line-of-sight) component of the velocity can be undetectably small. At intermediate angles, for example when the orbit of the planet-star system is tilted forty-five degrees to the line of sight, the radial velocity is proportionally diminished: in other words, the (unknown) orientation of the axis of the system would decrease all the measured radial velocities by a constant factor, which has the effect of making the planets appear less massive by that same factor. In those cases in which Earth lies close to the plane of the planet's orbit, the radial component of velocity is maximized. This means we can report only the minimum mass that the planet must have (that is, the mass calculated by assuming that Earth lies in the plane of the orbit). If we had both astrometric and radial velocity data on the same system, we could determine the tilt of the orbit and therefore calculate the mass of the planet accurately.

A third technique for detection of extrasolar planets is to search for light from the planet. At visible wavelengths, planets are enormously fainter than the stars around which they revolve, often by a factor of a trillion. Scattered visible light from the central star completely swamps the tiny, faint image of the planet. The relative brightness of the planet is actually highest in the infrared portion of the spectrum. Large, hot, distant planets are easiest to see. For planets whose "surface" temperatures are determined by the heat received from their parent star, the categories "hot" and "distant" are mutually contradictory. Planets large enough to have substantial internal heat sources, especially super-Jovian planets, are therefore favored.

The first direct detection of a planet of another star was reported in June 1998 by the University of Arizona's NICMOS camera on the Hubble Space Telescope. NICMOS caught a Jovian planet in the act of being ejected from a double-star system, trailing a long streamer of glowing gas. The great separation of the planet from its parent stars, well over 1000 AU, made imaging of the faint planet possible.

The fourth detection technique is to search for partial eclipses of the disk of a star by the passage of a planet across its face. Here again, as with the radial velocity technique, low inclinations of the orbit are favored — but in this case, the planet is usually not detected at all if the Solar System lies more than a degree or two out of the equatorial plane of the star. Thus this technique must miss the overwhelming majority of all planets. It is also subject to interference due to large variations in the brightness of stars caused by flares, sunspots, and other stellar "weather." This technique clearly favors planets that have a very large size relative to their star, and that orbit very close to the star. Ideally, one would search for super-Jovian planets orbiting small, faint M-class stars. But these stars are the most erratic in brightness of all main sequence stars, so variable that eclipses might not be detectable.

Despite all these difficulties, the radial velocity technique has so far found at least a dozen giant planets around other stars. The first, discovered by Michel Mayor and Didier Queloz of the Geneva Observatory in Switzerland, orbits 51 Pegasi, a yellow Sunlike (class G3) main sequence star 15.4 parsecs from the Solar System. The minimum mass of the planet, 51 Peg B, is 0.47 Jupiters. Incredibly, the planet follows a circular orbit only 0.05 AU from the star with an orbital period of 4.23 days. Jupiter, in contrast, is over one hundred times as far away from the Sun! Heating by the star must keep the "surface" of the planet at a temperature of about 1300 K (1800°F). These results were so odd that many astronomers have sought vigorously for possible loopholes in the data or in its interpretation. Nonetheless, the detection seems valid.

The credibility of the 51 Pegasi report has been enhanced by several other discoveries of close, massive planets. The white F7 star τ (tau) Bootis, for example, has a planet with a minimum mass of 3.87 Jupiters orbiting at 0.046 AU from its star with a 3.313-day period. The eccentricity of its orbit is small, about 0.018. This planet, excitingly named τ Boo B, must be at a temperature of at least 1600 K, even hotter than 51 Peg B. The white F7 star υ Andromedae also has a planet with minimum mass 0.68 Jupiters, orbiting 0.058 AU from the star. The orbital period is 4.61 days and its eccentricity is 0.15. This planet also is very hot, near 1400 K. The yellow G8 star 55 Cancri has a planet with a minimum mass of 0.84 Jupiters orbiting at a distance of 0.11 AU with a period of 14.65 days. The eccentricity of the orbit is 0.051. Its temperature must be close to 800 K. A second possible planet has also been reported orbiting this star: 55 Cnc C with a mass of at least 5 Jupiters, at a more respectable distance of 3.8 AU, with a long (but as yet poorly defined) orbital period of at least eight years.

The yellow G0 star ρ (rho) Coronae Borealis has a planetary companion with a mass of at least 1.1 Jupiters orbiting every 39.6 days at 0.23 AU. The F9 yellow-white star HD 114762 of at least 10 Jupiter masses orbiting at 0.38 AU (essentially the same as Mercury's distance from the Sun) with an 84-day period and a surprisingly high eccentricity of 0.25. The G4 yellow star 70 Virginis has a planet with at least 6.6 Jovian masses at a distance of 0.45 AU and with an orbital period of 116.6 days. The orbital eccentricity is an even more remarkable 0.40.

Even odder by our familiar standards of the proper behavior of giant planets is that of the yellow G2 star 16 Cygni. It has a planet of 1.5 Jupiter masses that orbits 1.7 AU from the star with a period of 2.19 years. But the eccentricity of its orbit is an astonishing 0.67! Then there is the G0 star 47 Ursae Majoris, whose planet weighs in at 2.8 Jupiters or more and orbits every 2.98 years at a distance of 2.1 AU. Its orbital eccentricity is a modest 0.03. Finally, the red M4 star Gliese 876 has a companion with over 2.1

Jupiter masses in a near-circular orbit at a distance of 0.21 AU, orbiting in 61 days, and the K0 star 14 Herculis was found to have a 3.3 Jupiter mass companion orbiting every 4.44 years at a distance of 2.5 AU.

Nonstellar bodies with even higher masses are also found: those that are significantly larger in mass than Jupiter, but too small to sustain hydrogen fusion reactions in their cores, are called *brown dwarfs*. The upper end of the brown dwarf mass range could be variously set at 0.01 to 0.08 solar masses, the former corresponding to the initiation of (temporary) deuterium burning, and the latter to sustained hydrogen fusion by the proton-proton chain. The latter is the more useful definition. The lower boundary of the brown dwarf region is, by comparison, harder to define. One criterion follows from the fact, established by detailed computer modeling of the structures of super-Jovian bodies with ages of several billion years, that there is a maximum radius for solar-composition bodies at about 2 Jupiter masses (0.0028 solar masses). Beyond that point, adding further mass causes so much compression in the deep interior of the body that its radius actually shrinks as mass is added. A better approach is to define brown dwarfs as the bodies for which deuterium burning, but not prolonged hydrogen burning, occurs. Then the low end of the brown dwarf region would lie near 0.01 solar masses (13 Jupiter masses).

Since bodies of a few Jupiter masses may maintain "surface" temperatures in the liquid water range in their upper tropospheres even at very great distance from a star, it is natural to wonder whether even an "orphan" super-jovian planet, one that travels through space without a star, might be a suitable home to life. The brown dwarfs, however, present their own distinctive array of problems and opportunities. They demand our immediate attention both because of their intrinsic interest, and because many brown dwarfs have recently been discovered in orbit around nearby stars.

Perhaps the central problem posed by these discoveries concerns the locations of giant planets and brown dwarfs: how can they form so close to a star? More than half of all the known bodies in these categories lie within 1 AU of their primary! Seven lie within 0.25 AU of their star, three between 0.25 and 0.50 AU, five between 0.50 and 1.00 AU, two between 1.00 and 2.00 AU, three between 2.00 and 4.00 AU, one (Jupiter) between 4.00 and 8.00 AU, one (Saturn) between 8.00 and 16.00 AU, and one at a greater distance. There appear to be two quite independent ways of making these large planetary bodies, each with its own preferred distance range. These mechanisms involve gravitational capture of gases and rotational splitting of a dense nebula or protostar.

The method favored for making the giant planets in our Solar System is gravitational capture of gases in the pre-solar nebula onto massive, Mars-size cores of dirty ice. This mechanism cannot make giant planets closer to the Sun than the point in the nebula at which ice condenses, since it is ice that provides most of the mass of the solid core. The G8 star 55 Cancri has a super-Jovian planets at distances of about 4 AU respectively, which may plausibly be attributed to the same origin as our Jovian planets because the ice condensation threshold lies much closer to these cooler, fainter stars. The brown dwarf companions of the orange K2 star HD18445 and the K4 star HD217580 also lie almost exactly on the expected ice condensation front, and the brown dwarf orbiting Gliese 229 lies far outside that front. But almost all the other known giant planets and brown dwarfs are too warm — some much too warm — for ice to be plausible.

A modification of this first method of formation is possible if the content of rocky material in the prestellar nebula was several times higher than in our own Solar Nebula. Then rocky bodies might grow large enough within the nebula to capture gases even at temperatures somewhat too high for ice condensation. But, as we have already seen, it is hard to push this process very close to a star: the ability of a solid body to capture gases drops off exponentially with temperature, and the ability of a star's tidal forces to strip atmosphere off a growing planet also increases dramatically close to the star.

Also, a very young giant planet may be a hundred times as large in diameter as the Jovian planet it will eventually become. Friction between this protoplanet and the nebula in which it is embedded may cause the planet to slow down in its orbital motion and spiral slowly inward toward its star. According to calculations by several authors, under some circumstances such radial migration may account for the appearance of a Jovian planet in close proximity to its star.

It would be nice if all brown dwarfs were very close to their parent star, and all Jovian planets were far away. Then we could plead the case that each class is made by its own distinctive mechanism. Unfortunately, as we shall see in detail in the next chapter, the bodies discovered to date follow no such trend. The only obvious pattern seems to be that the three examples of large companions around F stars are all close (0.046 to 0.38 AU) and one of the only two M stars with known massive companions is

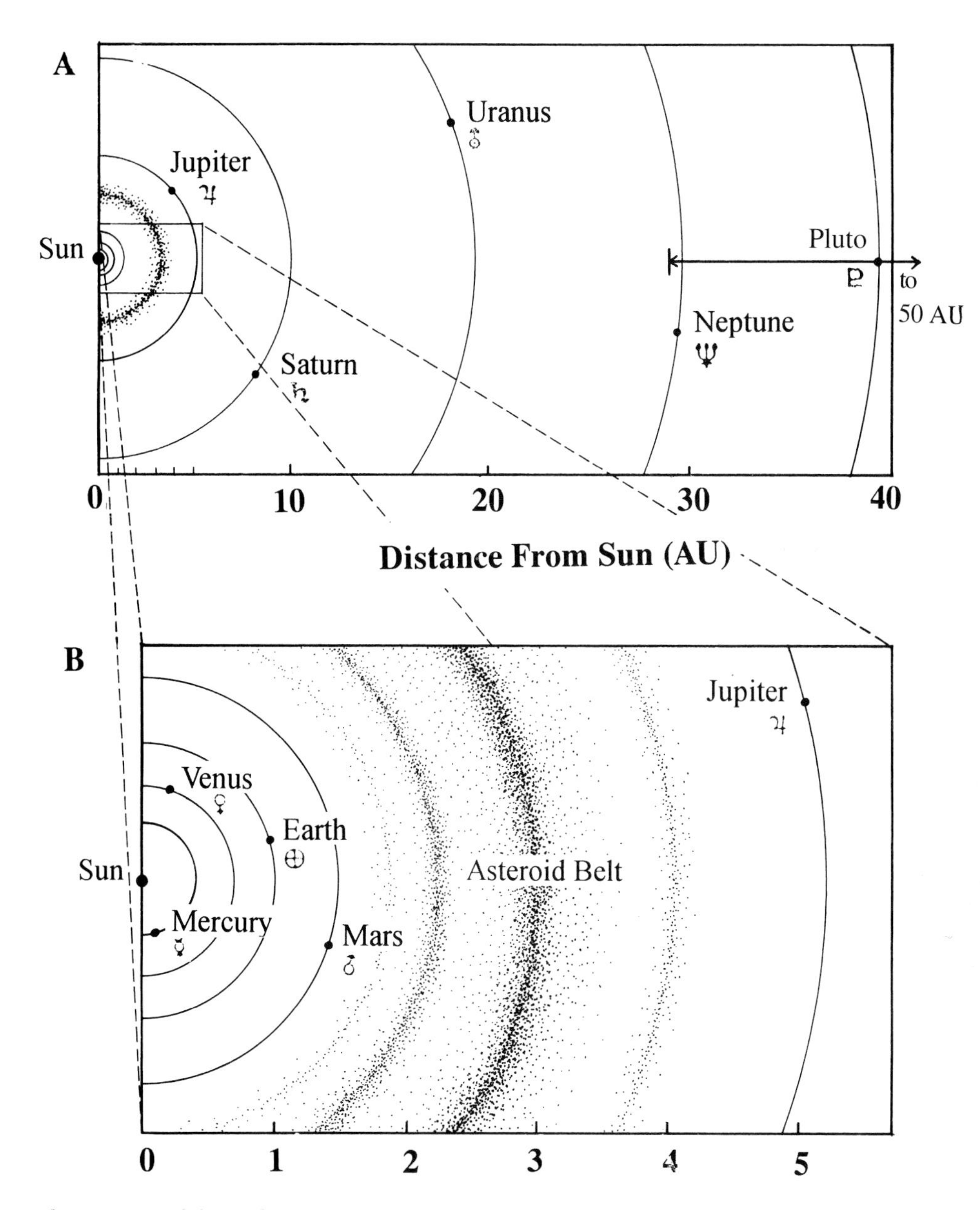

The Layout of the Solar System. Distances are given in units of the Earth's distance from the Sun (Astronomical Units, or AU). Part A emphasizes the outer planets. Pluto's eccentric orbit ranges from inside Neptune to as far as 50 AU from the Sun. Part B contains a detailed view of the inner regions, from the Sun to Jupiter.

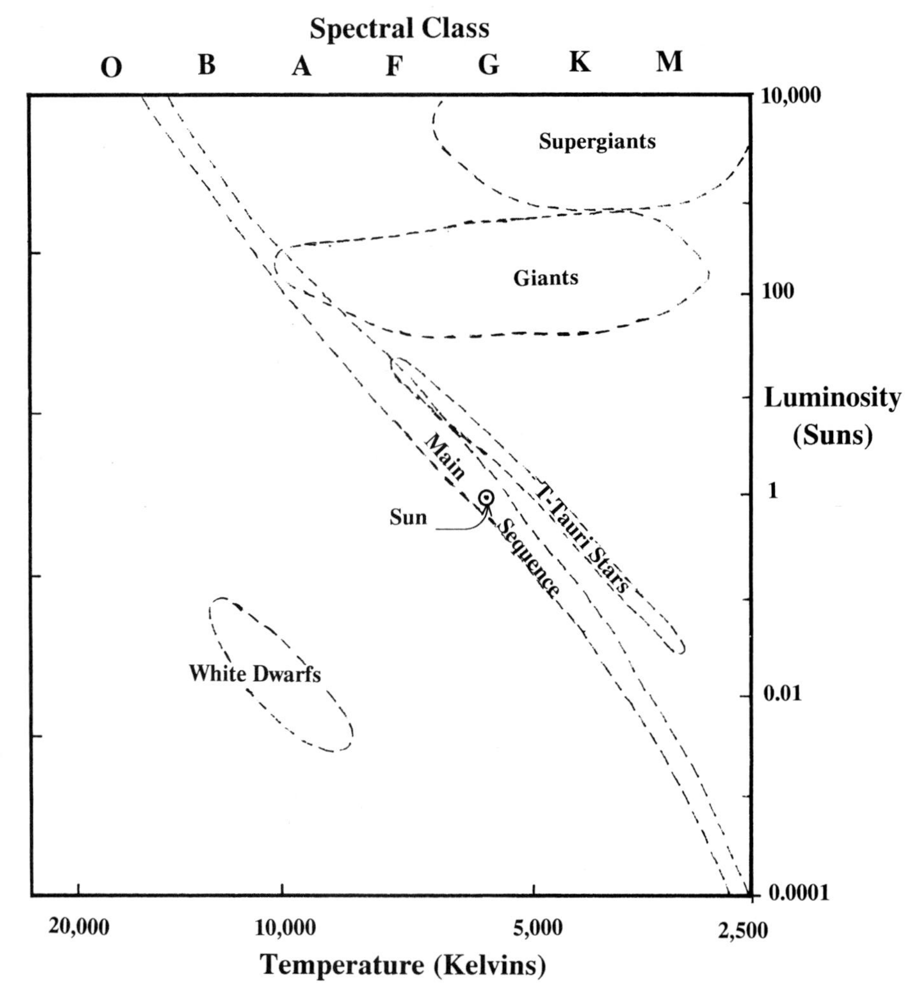

The H-R Diagram. The colors, temperatures, and spectral classes of stars are plotted against the luminosity of the star for the Main Sequence, white dwarfs, young T-Tauri stars, giants, and supergiants. Luminosities are relative to the luminosity of the Sun.

A tidally distorted, rotationally-locked-on inner world. Such a fate must be common for Mercuries. Such bodies, even if very large, must be rotationally locked onto their primary stars, in 1:1 spin-orbit resonances like the Moon or Jupiter's Galilean satellites, in a 3:2 resonance like Mercury, or conceivably a 2:1, 5:2, or higher resonance, if the eccentricity of the planet's orbit is high enough.

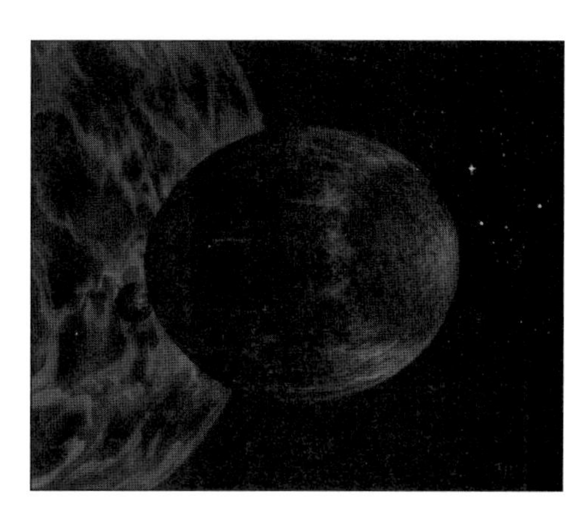

View from an Earthlike world orbiting a brown dwarf with a ring and smaller moons. The lighting reveals the presence of an unseen star about which this system orbits. Such a star can stabilize conditions on the Earthlike body, protecting it from the cooling of the brown dwarf.

A comparison of (left to right) a G-type and M-type MS star (with a brilliant flare in progress), an Earthlike planet, a Uranus-like planet, a Jovian planet, and a super-jovian planet (close to the mass at which gas giant planets achieve their maximum possible radius).

Worlds about a brown dwarf, as seen from a hypothetical world in orbit about the brown dwarf. A distant G-type star illuminates the system.

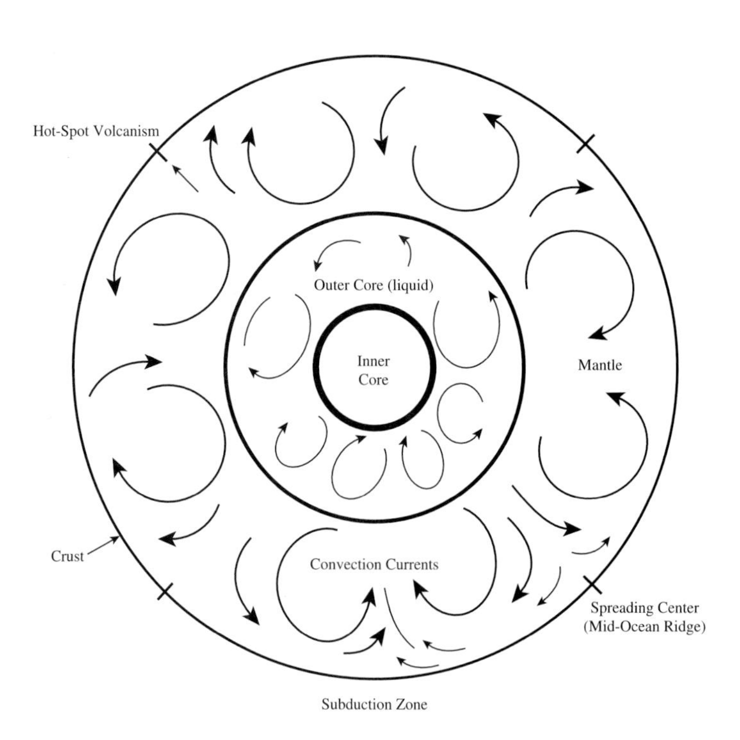

Structure, convection and plate tectonics of Earth. The inner core, probably an alloy of iron and nickel, is solid. The outer core is a liquid containing iron, nickel, sulfur, and probably silicon and oxygen. The mantle, composed of dense silicate and oxide rocks rich in magnesium and iron, is a hot, soft solid that convects slowly. The cold, brittle lithosphere includes the crust, which makes up only 0.4% of the mass of Earth.

An Earthlet; a warm, wet, Mars-sized world. Shallow seas cover about half its surface.

An Earthlike world orbiting the superjovian planet 16 Cyg B. A Sunlike G2 star illuminates both worlds. Visitors from another star system, traveling inside a hollowed-out asteroid, view the alien scene.

A view inside the atmosphere of a brown dwarf in its temperate phase. Lightning plays in and about tenuous water clouds in a blue sky, while the red heat of the deep interior leaks out from below. The gravitational acceleration in this region is close to 100 Earth gravities (an astonishing 1000 meters per second per second). After one second of fall, a dense body would be traveling several times the speed of sound.

View of Jupiter from the frozen surface of Europa. The biological promise of Europa's internal ocean is hidden beneath a thick layer of floating ice.

A globular cluster, home to perhaps a million stars and uncounted worlds.

Young worlds grow out of the debris disk surrounding a new T-Tauri star. A constant, intense bombardment by comets and asteroids sears the surface of the still-growing planets. When the sound and fury of accretion die down, what kinds of worlds will they be?

distant (44+ AU), but this apparent trend may be more due to the paucity of evidence than any profound physical controlling process. We should not, however, neglect the evidence from double and multiple stars, which apparently form by some kind of rotational fission of a prestellar cloud: these stars seem to have no particular preferred distance, running the gamut from contact binaries to bodies hundreds and even thousands of AU apart. Brown dwarfs and super-Jovian planets formed by rotational fission could be at any distance.

A final piece of evidence comes from studies of the rotation rates of main sequence stars. The most luminous, hottest, most massive MS stars, the members of the O, B, and A spectral classes, all show spectroscopic evidence of very rapid rotation. The rotation rate drops precipitously toward the bottom end of the main sequence: O stars usually have equatorial rotation speeds of about five hundred kilometers per second; F stars rotate at about fifty kilometers per second, and M stars rotate at speeds as low as five kilometers per second. In the Solar System, the Sun accounts for 99.9 percent of the total mass, but only 0.5 percent of the angular momentum! These observations are sometimes interpreted as evidence that long-lived stars can be substantially despun by means of their solar wind emission. Others argue that angular momentum is pumped outward by stars into their surrounding nebula, and thence into planetary systems. If the latter interpretation is correct, then F, G, K, and M stars must generally be accompanied by planetary systems. But if the solar wind despinning mechanism is correct, we would simply conclude that the G, K, and M stars spin so slowly because they are older and have had far more time to despin. Thus the significance of the rotation rates of stars remains unclear.

10

BROWN DWARFS AND THEIR CLOSE KIN

RED DWARF STARS, ALIAS M-CLASS main sequence stars, are the smallest bodies that can maintain fusion reactions in their cores. The brightest of them (subclass M0) have about half the mass and 4 percent of the luminosity of the Sun. Each successive subclass, such as M1, is slightly less massive and slightly less luminous than the preceding. M2 stars, at 0.4 solar masses, have luminosities of 0.02 Suns, and so on, down through M5 stars (0.2 solar masses, 0.01 Suns luminosity) to about M8 (0.00001 Suns luminosity; 0.07 solar masses), the faintest and coolest (under 2500 K) stars known. Because they are so small and cool, they are very dense, some weighing in at one thousand times the density of water. The faintest stars are so cool, in fact, that the atoms in their atmospheres combine to make a wide variety of simple molecules, such as carbon monoxide, molecular nitrogen, and hydrogen, and a number of carbon compounds beginning with the C_2 and C_2H molecules. Bodies only slightly smaller in mass than the faintest M stars cannot sustain high enough core temperatures to fuse hydrogen to helium; but, like the Jovian planets, they can maintain faint luminosity for many billions of years by slow shrinkage, converting their own gravitational potential energy into heat. These bodies, called brown dwarfs, are so cool (under 2000 K) that they radiate mainly at infrared wavelengths. Their spectrum is dominated by absorption bands of a vast variety of simple molecules, making their spectra very complex and hard to interpret. They are cool enough so that clouds of metals and silicates can condense in their atmospheres. Such bodies, which we can no longer call stars, are very faint or virtually invisible to the eye. They are called "brown" not because they would look brown to the eye, but because their complex spectra would lend them a muddy color, not a pure red.

The range from about 0.01 to 0.07 solar masses is actually transitional: stars in this range get hot enough to initiate fusion reactions that "burn" deuterium, but fail to sustain hydrogen fusion at a detectable level. They are the stars that flunked out of basic training.

For convenience, we shall count as brown dwarfs all the bodies between 0.01 solar masses (about 13 Jupiters) and 0.07 solar masses (the smallest reliable M stars). We know little about how they form: early speculation suggested that they might be products of routine star-formation that simply failed to pass quality control and never lit up their nuclear fires. Another view holds that the processes that give rise to close binary stars might also give rise to pairings in which one (or even both) of the partners is substellar in mass. Theorists' attempts to explain the formation of such massive bodies in light of Jovian planet formation have tended to be pretty unconvincing. Theory in fact did not provide much advance guidance about where to find brown dwarfs.

Our Solar System contains no brown dwarfs. In a way, this is a good thing: something as massive as a brown dwarf, orbiting the Sun as close as Jupiter or Saturn, would wreak havoc on the orbits of the planets, pumping up their eccentricities, then kicking them out of the Solar System, dropping them into the Sun, causing collisions between them, or simply swallowing them up in catastrophic collisions. But the absence of any such body in our backyard means that we are exceptionally dependent upon theory to tell us how they are put together, what they look like, and how they behave. Now, as a theorist, I'm the last person to cast aspersions on theory. But let's be reasonable here: the purpose of theory is to make sense of observations; to deduce generalities about the laws of nature from facts gathered arduously in laboratories, at telescopes, and via spacecraft. Developing a detailed body of theory about a class of bodies that have never previously been observed is fraught with perils, perhaps the greatest of which is that the theorist will overlook some niggling factor that, unbeknown to him, is very important in this unfamiliar population. But theorists are not the only ones with problems: an observer may fail to realize that certain types of data are crucially important, or interpret certain observations incorrectly, because of inadequate familiarity with theory. Science is not a footrace between observers and theorists; it is a game of tennis. It is the interplay between theory and experiment, the repeated interaction and redirection, that makes the game exciting and that makes progress possible.

Fortunately, the rapidly developing tools of astronomy have made it possible in the last two years to discover massive but invisible bodies in orbit around other nearby stars. These techniques are very much the same as those used to search for Jovian planets orbiting stars. As we discussed in Chapter 9, the first two techniques for detecting such bodies depend on the fact that the bodies being sought are massive enough to have detectable effects on the motions of their parent stars. Picture a brown dwarf with a mass of 0.01 Suns in orbit around a red MS star with a mass of 0.5 Suns. In fact, the two bodies orbit around their common center of gravity, which is one-fiftieth of the way from the center of the star to the center of the brown dwarf. As we discussed in Chapter 9, if we watch the system from a distance, we see the star moving along a wavy line, the waviness due entirely to the gravitational effects of the brown dwarf. If we measure the position of the star accurately, then, after about one orbit of the dwarf, we will know the amplitude (size) of the oscillation and its period. This technique has historically been known as astrometry (literally, "star measuring").

Alternatively, rather than measuring the position of the star many times, we may measure its velocity. According to the Doppler effect, any body that approaches us compresses its spectrum to shorter wavelengths. Likewise, any body that recedes from us stretches out its spectrum to longer wavelengths. Any body that is following a circular or elliptical orbit will show a constant shift of the wavelengths of all its spectral lines back and forth to longer, then shorter, wavelengths. This technique, as mentioned in Chapter 9, is called the radial-velocity method. Each cycle defines one orbital period of the body. Note that there is an extreme case, in which the orbit is exactly face-on to the observer (that is, viewed from the direction of the pole of the orbit), in which there is no radial component of velocity. But it is still possible to detect that body by the astrometric technique.

We already have on the books about a dozen recently discovered brown dwarfs. These turn out to be a pretty bizarre bunch of bodies. In every case, we have radial-velocity data only, with no independent way of determining the orientation of the orbit we are looking at. Suppose we see a velocity cycle of a certain size: is this the signature of a low-mass body whose orbit takes it directly toward us and then directly away from us, giving the largest possible radial velocity to its parent star, or is it a much more massive body in an orbit almost perpendicular to the line of sight, producing in the star velocity changes that are mostly not in the line of sight, and therefore looking very feeble to us? Generally, in the absence of astrometric data we can put only a lower limit on the mass of the brown dwarf. Some bodies that appear to be brown dwarfs based on radial-velocity measurements might be very faint MS stars whose light is lost amid the radiation of the brighter, more massive star.

The entire zoo of brown dwarfs reported to date is as follows: First, there is one in orbit around the faint star HD283750. The star is an orange K2-type main sequence star with about 75 percent of the mass of the Sun and a luminosity of about 0.2 Suns. The brown dwarf (remember, it has not been seen directly) orbits its star with a period of 1.79 days, at a distance of about 0.04 AU. It has a mass of at least fifty Jupiters, making it a possible brown dwarf or minimal MS star.

The bright yellow G0 star HD98230, about 14 percent more massive and 50 percent more luminous than the Sun, has a dark companion with a mass of at least thirty-seven Jupiters orbiting it every 3.98 days, at a distance of 0.05 AU. Then there is HD112758, a K0 orange star with 80 percent of the mass and 40 percent of the luminosity of the Sun, accompanied by a companion with a minimum mass of thirty-five Jupiters, orbiting its star every 103.22 days at a distance of about 0.35 AU. Interestingly, the brown dwarf's motion around the star is not uniform, as it would be for circular motion. The orbit must have an eccentricity of about 0.16.

The yellow G0 star HD140913, with about 1.14 solar masses, has an unseen companion with a minimum mass of forty-six Jupiters. Its orbital period is 147.94 days, orbiting at a distance of 0.54 AU from its star with an eccentricity of 0.61. The yellow G1 star HD89707 (1.05 solar masses; 1.25 solar luminosities) has a companion with at least fifty-four Jupiter masses in a 198.25-day orbit, 0.69 AU from the star. The orbital eccentricity is an astonishing 0.95. Note that either of these bodies might plausibly be a brown dwarf or an extremely faint M-class star.

The K5 orange star BD-04782 (0.7 solar masses; 0.1 solar luminosity) has a companion of at least twenty-one Jupiter masses in a 240.92-day orbit, 0.7 AU from the star. The orbital eccentricity is 0.28. HD110833, a K3 orange star (0.75 solar masses; 0.2 solar luminosity) has a companion of at least seventeen Jupiter masses in a 270.04-day orbit 0.8 AU from its primary. It has an orbital eccentricity of 0.69. The K4 orange star

HD217580 has a brown dwarf companion of at least sixty Jupiter masses, orbiting in 1.24 years at a distance of 1.0 AU, with an eccentricity of 0.52. Also, the K2 orange star HD18445 (0.75 solar masses; 0.2 Suns luminosity) has a companion of at least thirty-nine Jupiter masses, an orbital period of 1.52 years, and a distance of 1.2 AU, with an eccentricity of 0.54. The G2 yellow solar-type star HD29587 has a dark companion of at least forty Jupiter masses in a circular orbit at 2.1 AU, with a period of 3.17 years. And finally, the red M1 star Gl 229 has a companion of thirty to fifty-five Jupiter masses in a 400-year orbit at a distance of approximately 44 AU. So little of its orbital cycle has been tracked that the eccentricity of the orbit is as yet unknown.

The radial-velocity results seem to span a relatively narrow mass range (seventeen to about seventy Jupiters), but we must again remind ourselves that the measured masses are lower limits on the true masses. It is remotely possible that many of these bodies are actually very faint MS companion stars in double-star systems, whose orbital poles are accidentally aimed more or less directly at Earth. From such a vantage point, we see very small radial velocities, and estimate masses down in the brown dwarf range. Statistically, it seems extremely improbable that the bodies we see are *all* red MS stars in unusual orientations. But, in the absence of astrometric data on these systems (which works wonderfully irrespective of the orientation of the axis of the orbit), we cannot claim certainty for any of the inferred masses.

Our putative collection of brown dwarfs exhibits a wide range of orbits. The very closest companions, those of HD283750 and HD98230, have very nearly circular orbits, as expected from the effect of powerful tidal interactions at such small distances. The next in order of distance, HD112758, has a moderate eccentricity of 0.16. The next six brown dwarfs in order of distance (HD140913, HD89707, BD-04782, HD110833, HD217580, and HD18445) all have very high eccentricities of 0.28 to 0.95, averaging 0.60. The next, that of HD29587, has a near-zero eccentricity. So far we know nothing of the orientation of the orbital planes of these bodies relative to the equators of their stars, or relative to the orbits of other bodies that may also be present around these stars.

In many ways these orbits remind us of double-star systems, in which very close companions are sometimes seen. They certainly do not remind us of the orbits of the giant planets in our own Solar System. In fact, none of them have orbital distances between 2.2 and 40 AU, where

Jupiter, Saturn, Uranus, and Neptune lie. Also, the presence of high eccentricities seems to indicate that these bodies represent a very different population from the gas-giant planets in our own Solar System. One can easily enough imagine a system containing several massive giant planets or brown dwarfs in orbits that get a little too close to each other for long-term stability. Gravitational interactions of the bodies could easily build up the eccentricity of one or more of the orbits, and even eject one or more planets from the system by giving them an eccentricity greater than 1.00. The presence of a brown dwarf in a highly eccentric orbit may offer evidence that other planets from that system have been forcibly ejected, sent out to roam in the interstellar darkness. We shall revisit these orphan bodies in Chapter 15.

The extreme closeness of some of these bodies to their parent star is also striking. It is clear from computer simulations of the evolution of a dense preplanetary gas and dust disk that forming planets tend to spiral inward. The reason is straightforward: a planet-size body is little affected by gas drag, and tends to travel around its star at orbital velocity. But the gas disk partially supports itself by its own internal gas pressure. It therefore moves at a lower speed than a planet at the same distance from the Sun. Planets therefore tend to plow through the gas disk and sweep up slower-moving matter. This drag force slows the planet and causes it to spiral inward toward the central star. Since preplanetary nebulae tend to have a well-defined inner edge, this inward evolution may continue until the planet runs out of resisting medium.

One direct result of these close orbits is that some of the brown dwarfs (for example, the first two on our list) must be maintained at very high temperatures by their primary stars even without regard to their internal heat sources. But brown dwarfs will more often be located far enough from their primaries that heating by the star would not have a significant effect on their temperatures. The internal heat flow of brown dwarfs must commonly provide "surface" temperatures of 500 to 2000 K, intermediate between giant planets and the coolest main sequence stars. The lower-mass brown dwarfs are about the same diameter as Jupiter (maximum diameter occurs at about two Jupiter masses). The most massive brown dwarfs will be about the size of Saturn—and about the size of an M9 star. Any brown dwarf hotter than about 750 K would emit a faint glow detectable to the human eye, even though most of the emitted radiation is at infrared wavelengths to which the eye is completely insensitive. A 700

K brown dwarf would be more easily detected by your hands (via infrared heat radiation) than by your eyes.

Very recently, in June 1998, the infrared Two-Micron All-Sky Survey (2MASS), headed by James W. Liebert of the University of Arizona, has discovered great numbers of very small stars, fainter and cooler than M9 MS stars. These bodies are so cool that they emit mostly in the infrared and are undetectably faint in visible light. Since none of these infrared stars in the 2MASS survey are in orbit around other stars, neither radial-velocity nor astrometric mass measurements are possible. The masses of these bodies are therefore not yet known, but, based on their luminosities and observations carried out on the Keck II telescope by Davy Kilpatrick of Cal Tech, about twenty are considered to be members of a new class of main sequence stars called L dwarfs, and six are tentatively identified as brown dwarfs. Such faint stars can be observed only if they are very close to the Sun. The fact that so many were seen in a random sample of 1 percent of the sky suggests to Kilpatrick that there may be about two thousand such infrared stars within fifty light years of the Sun. By this estimate, L dwarfs and brown dwarfs would be much more numerous than every other class of stars combined.

THE LARGE number of giant planets and brown dwarfs that have been found to be very close to their parent stars is, at first sight, very startling. Of the several hundred stars searched to date, about 3 percent have been found to have giant planets (one-third to 13 Jupiter masses) and about 1 percent have been found to have brown dwarf companions (13 to 90 Jupiter masses). But these observations have for the most part been conducted over just a few years: there has been insufficient time to discover bodies with orbital periods of more than a few years. Both the astrometric and the radial-velocity technique are best at finding massive planets that are close to their primaries, which again favors short orbital periods. There is as yet no reason to assume that the systems found so far are a representative sample of other planetary systems. The list of discoveries is very strongly skewed toward rare, highly detectable (massive, close) planets. The population of planets similar in size and distance to the giant planets in our Solar System will not be known — indeed, will not be knowable — for several years.

Brown dwarfs present a challenging but not necessarily impossible environment for the origin of life. The dense, convective lower atmospheres (tropospheres) of brown dwarfs will, without exception, be too warm for the presence of liquid water. Sterilizing temperatures are in many cases prevalent throughout the troposphere, and in every case are present only a few hundred kilometers below the coolest, most benign layers. Stratospheric temperatures in the liquid water range are conceivable for the very smallest brown dwarfs if they are located far from their star, but stratospheres are especially unfriendly locations for the development of life. Stratospheres are nearly uniform in temperature and virtually devoid of convective motions, so that there is no way to suspend and lift cloud particles and droplets. Particles rapidly fall through the base of the stratosphere into the upper fringe of the troposphere (the tropopause), where convection efficiently homogenizes the atmosphere and where the temperatures are high enough to evaporate icy condensates. If one takes the brown-dwarf range as 0.01 to 0.07 solar masses, then we can conclude that life forms chemically similar to those on Earth could not exist, due to the absence of environments with suitably moderate temperatures and good temperature stability. If, however, as some do, we place the dividing line between gas-giant planets and brown dwarfs at the maximum radius point (about two Jupiter masses), then the smallest bodies in this range would be similar to, and perhaps a little more congenial than, Jupiter and Saturn. These are the bodies we discussed in the previous chapter.

SINCE BROWN dwarfs can maintain temperatures and luminosities not much inferior to the feeblest main sequence stars, it is quite conceivable that they may provide the necessary heat for the maintenance of benign, stable temperatures on the surface of a satellite. After all, if stars can have planets and planets can have satellites, why couldn't a brown dwarf have its own retinue of planets? Two scenarios come to mind: a brown dwarf accompanied by its own world-sized moons, all in distant orbit around a stable MS star; and an "orphan" brown dwarf traveling with its own planetary system through interstellar space.

The principal difficulties faced by such scenarios are the steady cooling of the brown dwarf and the absence of ultraviolet radiation to serve as an energy source for the origin of organic matter. Brown dwarfs are far too cool and stable to generate ultraviolet light of their own by thermal emission. Simply put, solar material is a rich source of organic matter if and only if there is a source of high-energy light or particles to drive chemical reactions far away from low-temperature equilibrium. The energy-rich products of such reactions are the power source for prebiotic chemical evolution, the origin of life, and biological evolution.

Residents of Earth happen to think that UV is very important because our star emits a modest but adequate stream of ultraviolet photons, and we observe a host of important processes that are driven by that light. These include the production of organic matter on the giant planets and the largest satellites and the production and maintenance of Earth's ozone layer. Fortunately, there are other sources of energy besides ultraviolet light from stars for driving chemical reactions. For example, cosmic rays permeate the Galaxy. They constantly strike every body in the Solar System, although atmospheres, planetary magnetic fields, and especially the Sun's solar wind all moderate their effect: the solar wind effectively stands off the low-energy cosmic ray particles, which make up most of the population of cosmic rays coursing through the Galaxy. A brown dwarf would be incapable of mustering a decent defense against these particles, leaving itself and its planet-size satellites exposed to their chemical effects.

Second, convective planetary atmospheres always contain lightning. Lightning discharges are white-hot, commonly 20,000 to 30,000 K, as hot as the surfaces of O or B stars. Lightning discharges are therefore powerful, albeit brief, sources of ultraviolet light, and generators of long, cylindrical shock waves (thunder). The lower atmosphere of the brown dwarf itself is so warm that it is weakly conducting, which tends to short out electrical-charge accumulations before they can grow large enough to trigger lightning discharges. Besides, the atmosphere of the brown dwarf is both opaque enough to prevent the UV light from lightning discharges from getting out, and hot enough to destroy any interesting products very efficiently. But the worlds in orbit around the brown dwarf, if they have atmospheres of their own, can generate their own lightning. The ions produced by lightning discharges are potent drivers of chemical reactions, and the ultraviolet light tears apart methane, water vapor, and other simple molecules with abandon, making reactive fragments that will readily recombine to make larger, more complex molecules.

Third, radioactive-decay energy is omnipresent in the Universe. The heavy elements necessary for formation of planets are made and ejected by supernova explosions that also produce a great variety of radioactive nuclides with a wide range of half-lives. The surviving radioactive species after a few million years of planetary formation and shrinkage are principally isotopes of potassium, uranium, and thorium, with very minor traces of a dozen or so others. These radioactive elements are present in the rocky component of all planets. In some cases, significant amounts of potassium may be extracted into water. But no planet with temperatures in the liquid water range can have these elements in their atmospheres (except, of course, in the immediate aftermath of a nuclear war!). For radioactive decay to be effective in driving the chemical evolution of volatiles, the radioactive elements must be in direct physical contact with the volatiles. The planetary surface is indeed an interface between these two essential components, but far less than a millionth of all the radioactive atoms are close enough to the surface to actually have any effect on the volatiles. The best opportunity for radiolysis (chemical reactions driven by radioactive decay) is in potassium-bearing solutions such as oceans.

Fourth, a more indirect reflection of the same source of energy is the eruption of hot magmas onto the planetary surface. In most cases, the ultimate energy source for generating magmas is radioactive decay; however, other sources of energy, such as tidal interactions, may in some settings (close to a large gravitating body) be very important.

Fifth, a ubiquitous source of chemical energy in planetary systems is the impact of small bodies on planets. Cometary and asteroidal debris striking at high velocities generates powerful explosions with transient, very high temperatures, ultraviolet emission, and violent shock waves. Although we have no idea how much energy is brought in each year by impacts in other planetary systems, everything we know about the origin and growth of planetary systems suggests that this effect is essentially universal. (The utter absence of impacting bodies in a planetary system would create the presumption that some fanatically safety-minded, risk-intolerant civilization had gone to the trouble and expense of cleaning up Nature's mess.)

The emerging picture of a planet orbiting a brown dwarf is astonishingly un-Earthly. A warm, muddy red, barely glowing failed star looms enormous in the sky. Any planet close enough to be kept warm by this feeble furnace would have to be so close that it would be despun like the large moons of Jupiter, or at least captured into a low spin-orbit resonance like Mercury. The pace of chemistry — indeed, the pace of life itself — would be snail-like in this energy-poor environment. Planets of Earthlike composition with temperatures in the liquid-water range, or very water-rich bodies like the Galilean moons of Jupiter, would be kept warm by the heat emitted by the brown dwarf, but all their chemistry and biology would have to be driven by cosmic rays, lightning, impacts, and radioactive decay.

The search for life-friendly environments around brown dwarfs has led us from their interiors to their cloud tops, and thence to their satellites. The biological significance of bodies such as the Galilean satellites, whether in orbit around gas-giant planets, brown dwarfs, or main sequence stars, seems obvious and clearly deserving of further exploration. This we will undertake in Chapter 11.

PLANETS AROUND PLANETS

THE FOUR TERRESTRIAL PLANETS AND four gas-giant planets of our Solar System are, without doubt, a pitifully incomplete sample of the possible range of worlds. But we have eight other examples of planet-size bodies to instruct us, if we will only pay attention. These are Pluto, a pathetic wimp of a planet compared with its burly neighbors, and the seven largest planetary satellites. The latter seven include Earth's Moon, which we used as an object lesson in the chapter on Mercury-like planets; the four Galilean satellites of Jupiter (Io, Europa, Ganymede, and Callisto); Saturn's sole large satellite, Titan; and Neptune's sole large satellite, Triton. All of these large moons are at least three thousand kilometers in diameter. No other moon in the planetary system has more than one-fifth the mass of the smallest of these large moons. Uranus and Pluto have only compact and midsize moons, Mercury and Venus have none, and Mars has only two tiny satellites.

The family of four large satellites orbiting Jupiter was discovered by Galileo in 1609, shortly after the first astronomical use of the telescope. The four were found to lie in Jupiter's equatorial plane, in concentric circular orbits, with orbital periods of 1.77, 3.55, 7.15, and 16.69 Earth days, respectively. Galileo named them after four characters associated with Jupiter in classical mythology.

The Jupiter system, with four small inner satellites, four large Galilean satellites, two outer satellite families of four members each, and even a narrow, faint ring, is complex enough to inspire comparisons with the Solar System itself. But, upon close inspection, much of what we see there is new to our experience. The great mass of Jupiter and the relatively intimate scale of the Galilean system provide new insights into the range of possible planetary behavior.

Io, the innermost of the four Galilean satellites, orbits 422,000 kilometers from Jupiter's center, just a little more than the distance of the Moon from Earth. Because Jupiter is so large, Io is actually closer to the top of Jupiter's atmosphere than the Moon is to the surface of Earth. Because Jupiter has eleven times the radius of Earth, it looms enormous in the sky as seen from its large moons: its disk as seen from Io is 19.5 degrees in diameter, compared to 0.5 degrees for the Moon as seen from Earth. The area of Jupiter's disk (allowing for the flattening of the planet caused by its rapid rotation) is some 280 square degrees, compared with 0.2 square degrees for the Moon in Earth's sky. Although Jupiter covers 1400 times as much sky, it is not quite that much brighter than the Moon: the intensity of sunlight on Jupiter is less than that striking the Moon by a factor of $5.2 \times 5.2 = 27$, because Jupiter is 5.2 times as far from the Sun. This is partly offset by the fact that Jupiter's bright clouds reflect an impressive 44 percent of the incident sunlight, whereas the Moon is covered by a dark rock powder that reflects only 6 percent of the sunlight that strikes it. Allowing for both effects, the full Jupiter as seen from Io is 380 times as bright as the full Moon seen from Earth, a truly stunning sight.

Since Io always keeps the same side facing Jupiter (a 1:1 spin-orbit resonance), an observer anywhere on Io would see Jupiter nearly stationary in the sky. Slight departures of the orbit from perfect circularity and from perfect alignment with Jupiter's equatorial plane would actually cause Jupiter to execute a small circle in the sky, but that would not be obvious to a casual observer.

Jupiter's powerful magnetic field has trapped high-energy protons and electrons from the solar wind. These trapped particles, held in radiation belts in Jupiter's inner magnetosphere, travel around Jupiter in rough synchrony with its nine-hour-and-fifty-minute rotation. The heart of the radiation belts lies inside Io's orbit. Jupiter rotates about 4.3 times for each trip of Io around the planet, so the trapped particles are driven against the trailing side of Io by Jupiter's rotational motion. The intense radiation from these million-volt particles makes Io's surface one of the deadliest places in the Solar System. However, the bombarding particles have only limited ability to penetrate solids: at a depth of only a few meters in Io's crust, the fierce blizzard of protons and electrons would be quite undetectable. Also, the intensity of radiation is much less on the leading side of Io, where the radiation belt particles overtaking the satellite are screened out by Io itself.

Io's surface is completely covered with volcanic landforms. At any moment, several major volcanic eruptions are in progress, some spraying gas and dust to heights of 250 to 300 kilometers. But even the sites of the most violent eruptions have little resemblance to towering Martian or terrestrial volcanoes: the erupted materials on Io are weak and fluid, forming long, sinuous "lava flows" that run out to great distances from the low volcanic vents. Some mountain ranges are seen, possibly silicate mountains mantled by thin coatings of volcanic sulfur, but most surface features seem to be made of weak, easily melted sulfur. Impact craters are absent: the surface is so young, so frequently recycled, that craters are negligible. Most of the surface appears to be made of vast flows of sulfur that erupted as a liquid and solidified in thin layers. The principal gas driving the volcanic eruptions is sulfur dioxide. Deposits of white sulfur dioxide "snow" are seen locally. Water, usually the most common volatile in solid satellites and planets, is extremely rare or absent.

Gases ejected from Io, including sodium vapor and sulfur and oxygen compounds, form an extended torus or doughnut following Io's orbit around Jupiter. Under the impact of energetic particles and solar ultraviolet radiation, the gases in this torus are ionized to make a thin, glowing plasma.

The principal heat source powering Io's volcanic activity is tidal interaction with the other large Galilean satellites and Jupiter itself: the satellites perturb each other's orbits, making them slightly eccentric. But the satellites then interact with Jupiter to damp out the eccentricities of the orbits. The energy dissipated by tidal distortion of their interiors appears as heat. In the case of Io, the Galilean satellite closest to Jupiter, the rate of heat generation by tidal forces and by induction heating caused by passage through Jupiter's inner magnetosphere is high enough to maintain a constant state of violent eruption. The same factors heat the other Galilean satellites to different degrees: although these forces are less severe on Europa, they are nonetheless still important there.

The simplest explanation for the oddities of Io seems to be that the body once contained water, but that violent volcanism and constant recycling of its crust has allowed the escape of all light gases, especially hydrogen. Oxidation of the crust of Io by extinct water has converted primordial sulfides into a mixture of elemental sulfur and sulfur oxides, which remain today because they are too heavy to escape. Sulfur, being very easy to melt and having a low density, is preferentially erupted when the crust is heated.

Io is a world bizarre enough to merit a book of its own; but, given our present overriding interest in places suitable for the origin and evolution of life, we must confess disappointment in our summary treatment of this entity and move on.

EUROPA IS also a strange world, a convincing argument that terrestrial chauvinism is unwarranted and misleading in the Universe at large. As utterly un-Earthly as Europa appears, it is the most likely body in the Solar System to harbor native alien life.

As the second of Jupiter's moons, Europa is subject to less bombardment by the radiation belts than Io, and much less intense induction heating by Jupiter's magnetic field. And as the smallest of the large Solar System satellites, it was reasonable to expect that Europa would be a relatively cold, inactive body.

Europa is also obviously different in appearance from Io. In contrast to Io's yellow-orange craterless surface, mountain ranges, violent volcanic activity, immense sulfur lava flows, and black and white splotches, Europa is clearly an ice-covered world. Its surface is delicately tinged a pale yellow, apparently by sulfur blown out of volcanoes on Io and into orbit around Jupiter, but the infrared spectrum of the surface is dominated by the distinctive fingerprint of the absorption bands of water ice. The surface, as shown first by the Voyager missions and more recently stunningly revealed by the Galileo orbiter, looks like pack ice on a shallow ocean on Earth, lightly sprinkled with small impact craters. It is obvious from the Galileo images that the surface is made up of thin, mobile ice rafts that collide, fracture, override each other, weld together, and split up in a prolonged and complex choreography. But there is no way to date the crustal activity we see: it is conceivable that all that activity has ceased, leaving a frozen-up fossil landscape that preserves a record of ancient crustal mobility.

The density of Europa suggests that it is made of a mixture of rock and ice, but not close to solar proportions: a surface ice layer tens of kilometers thick is about all that can be reconciled with the overall density. Conduction of heat through a thick ice layer is so inefficient that a large temperature difference builds up across the insulating ice layer. When I looked at this problem for the first time, back in 1971, I realized that eventually temperatures at the base of the ice layer get high enough so that melting would begin. With a large enough internal heat source, most of the ice melts to produce a deep ocean with a thin layer of cold Ice I floating on top, moving about at the whim of the ocean currents. Freezing of water into dense, high-pressure forms of ice is not very important on Europa because of the low acceleration of gravity (one-seventh that of Earth): at the base of a fifty-kilometer layer of water the pressure would be only seven hundred atmospheres, similar to that at the bottom of Earth's deepest trenches and well below the two-thousand-atmosphere minimum pressure for appearance of high-density ice (Ice III). I recall joking with my undergraduate planetary sciences class at MIT in the early 1970s that Europa's ocean would be a great place for whales. The problem, of course, was where to get their minimum daily requirement of krill.

Tidal pumping of Europa actually puts energy into the silicate mantle well below the ocean floor, not directly into the crust. That energy must, as on Earth or the Earthissimo developed in Chapter 8, make its way upward to the vicinity of the ocean floor. There, infiltration of cracks in the oceanic crust by seawater will inevitably generate great volumes of superheated water analogous to Earth's ocean-floor hydrothermal vents, the "black smokers." On Earth, we find complex ecosystems surrounding these vents, containing many species never seen anywhere else on Earth. They feed off the current of hot, nutrient-laden water that gushes out of the black smokers into frigid ocean-bottom water. This ecology lives on Earth, but is independent of the Sun. It has no need for photosynthesis, for molecules produced by ultraviolet light, for solar heat, or for a day-night cycle. It is effectively another, alien world.

The idea of extremely deep oceans on Europa, complete with central heating and energy sources for making food, inspired Arthur C. Clarke to take the ball and run. In his 1982 book, *2010: Odyssey Two*, Clarke postulated the discovery of native oceanic life by an expedition to Europa. (He also populates Jupiter's atmosphere with a wide variety of life forms.)

How might grossly Earthlike life arise on Europa? The principal source of high-energy molecules must be the ocean-floor hydrothermal vents. Let us neglect other minor sources of energy, such as radioactive decay of potassium-40 dissolved in the ocean, and concentrate on this one mechanism. Hot magmas, partially melted rock intruding tens of kilometers below the ocean floor, will generate a mixture of carbon monoxide, nitrogen, water vapor, and traces of hydrogen cyanide, formaldehyde, and other simple organic building blocks. As the hot magmatic gases rise through the crust, they encounter and mix with cold ocean water at pressures of about one thousand atmospheres. At these high pressures and moderate temperatures, water vapor and carbon monoxide spontaneously react to make carbon dioxide and hydrogen. Nitrogen also reacts with hydrogen to make ammonia, and some carbon monoxide and carbon dioxide get converted by hydrogen into a mixture containing methane and more water vapor. In the process, more hydrogen cyanide and formaldehyde get made. Further cooling converts carbon dioxide, ammonia, and water vapor into ammonium bicarbonate and ammonium carbamate. Hydrogen cyanide can begin to polymerize into a variety of reactive nitrogen-containing organic molecules, including the organic base adenine. Hydrolysis of hydrogen cyanide (reaction with water) also makes amino acids, which readily link together in chains to make peptides and proteins. Formaldehyde polymerizes to make simple sugars. Sugars join up pairwise with organic bases to make nucleosides. Phosphates dissolved out of the rocks by hot water link up nucleosides into long chains—DNA and RNA. The race to life is off and running. And this scheme is self-contained, independent of the Sun.

The idea of such well-hidden life is startling, but actually demonstrated by two separate types of life forms on Earth. Quite independent of the ocean-vent communities, living microorganisms have been found at considerable depths in the crust, living in hot, moist cracks in rock and sustained by organic and even inorganic materials dissolved in the water. It is, however, unclear whether these communities are truly independent of surface life. Even those that receive no essential organic nutrients washed down from the densely populated surface in groundwater are genetically terrestrial: they are highly specialized descendants of normal Earth life adapted to an extremely hostile environment. If life ever prospered on the surface of Mars, its similarly adapted descendants may still live deep within the planet's crust. Life, once established on a planet, is immensely flexible and innovative.

GANYMEDE, MUCH farther from the center of the Jovian system, is important more as a disturber of the motions of Europa and Io than as a victim of their effects. It is so far out that magnetic induction heating is negligible, and so much more massive than Europa and Io that these two smaller satellites have little effect upon Ganymede's history. Ganymede is nearly the size of Mars. The surface of Ganymede has many grooves and cracks reminiscent of Europa, but they are far less prominent and far more heavily overlaid by craters. Evidently they reflect an earlier time when a relatively thin, mobile crust covered Ganymede. The newly formed Galilean satellites, heated by the high luminosity of Jupiter during its rapid early contraction, may even have been warm enough for liquid water to be stable on their surfaces.

The oceans of Ganymede are vastly more massive than those of Europa, perhaps six hundred to eight hundred kilometers deep. The pressure at the base of the ocean is far into the stability range of high-density forms of ice. Deep, cold ice, Ice VI, very likely coats the silicate interior. It would not be surprising to find that the entire ocean of Ganymede has frozen in response to the cooling and contraction of Jupiter, leaving a relatively thin low-density crust of Ice I resting upon consecutively deeper, denser layers of Ice III, V, and VI. Ganymede, if not completely dead, is certainly far less lively than it was in its heyday.

CALLISTO, THE most distant of the Galilean quartet, does not share in the near-resonant interactions of the other three Galilean satellites. Callisto is similar in composition and intrinsic density to Ganymede, but is somewhat smaller and colder, a more extreme version of the same long-frozen scheme. But the spacecraft images of Callisto put the case even more strongly: there is virtually no trace of an early, active phase with rifting and drifting of a thin ice crust. Instead, the entire surface is densely covered by small craters. Large craters are conspicuous by their absence: it appears that ice is weak enough to slump and wipe out the largest crater features over billions of years, leaving only the vastly more numerous and more recent smaller craters to attest to the long bombardment history. The very high crater density attests to the great age of the surface of Callisto. If Ganymede is a probable corpse, then Callisto is a certain one.

On both Ganymede and Callisto we find a number of crater chains of roughly evenly spaced, equal-sized impact craters, often with two or three dozen craters in the chain. These are the scars of impacts of comet showers formed by the tidal disintegration of large comets that have passed too closely to Jupiter for their own good and have been torn apart by Jupiter's tidal force into a spray of fragments. If the famous Jupiter-impacting "comet" Shoemaker-Levy 9 had missed Jupiter and hit one of the Galilean moons instead, it would have left just such a scar. This and other major impact events are discussed in more detail in my book *Rain of Iron and Ice*.

SATURN HAS a diverse retinue of satellites, of which one, Titan, is the most massive by more than a factor of fifty. Titan is about the size and density of Ganymede: Titan, Ganymede, and Callisto are all intermediate in size between Mercury and Mars, and about ten times the mass of Pluto. If the gas-giant planets were to vanish, the satellites they leave behind would surely be counted as planets in their own right.

Titan, however, sets itself apart from the Galilean satellites by having a dense, cloudy atmosphere. A reddish haze or smog completely obscures the ground at visible wavelengths, although radio and radar waves can pierce the atmosphere to reveal the general layout of the surface.

The atmospheric pressure at the surface of Titan is about 1.5 atmospheres, surpassed only by Venus among the nongiant planets. The surface temperature is only 94 K. The atmosphere is principally composed of nitrogen containing some percent of methane. Traces of hydrogen, carbon monoxide, ethane, hydrogen cyanide, cyanogen, carbon dioxide, and several other relatively simple compounds of carbon, nitrogen, and hydrogen are also seen. The organic chemistry of Titan's atmosphere is driven by the absorption of ultraviolet light from the Sun and by the impact of protons and electrons from Saturn's radiation belts. The surface of Titan is much too cold to allow detectable amounts of either ammonia or water vapor. The oxygen compounds seen must be descendants of vapors from the evaporation of icy and rocky meteors that burned up in the atmosphere.

It is fairly easy to estimate the rate at which methane in Titan's atmosphere is being destroyed by solar ultraviolet radiation. The entire atmospheric content of methane should be destroyed in a few tens of millions of years. There must therefore be a massive reservoir of methane in the crust or on the surface of Titan that evaporates to maintain the atmospheric methane content and compensate for its photolysis. Even more striking is the result of exposing the present Titan atmosphere to solar radiation for billions of years: the cumulative mass of photolysis products reaches tens of kilograms of heavy hydrocarbons per square centimeter of surface area of Titan. This is enough to cover the entire surface of Titan in a layer of lighter fluid and gasoline several hundred meters deep. There was some suspicion that, if Titan had a very flat surface, the entire crust might be inundated by a global ocean of hydrocarbons. But the radio data show that the surface is far from featureless. The heavy hydrocarbons must collect in seas rather than in a global ocean.

The very low surface temperatures on Titan are compatible with the presence of lakes or oceans of liquid nitrogen with dissolved methane and other organic constituents. Such a liquid is of very little biological significance. The surface is far too cold for any liquid based on ammonia or water to exist. Titan would be vastly more interesting if it had a surface temperature of 200 K or higher, so that ammonia-water solutions could exist. A massive greenhouse effect would be required to achieve such temperatures (more than doubling the present surface temperature). This might conceivably be accomplished if ammonia were added to the atmosphere in large quantities; unfortunately, however, the surface of Titan is presently so cold that this does not seem to be possible.

Triton, the largest satellite of Neptune, looks very much like a smaller, colder variation of Titan. The *Voyager 2* flyby of Triton in 1989 found a number of strange surface streaks and two plume-like features that appear to be geysers of liquid nitrogen. A tenuous atmosphere mainly composed of nitrogen lightly coats its surface, which has a mean temperature of only 38 K. Seasonal nitrogen snow is seen in the winter hemisphere. Triton's lightly cratered surface appears to have been extensively resurfaced at some uncertain time in the past. It is possible that the heating event was caused by the capture of Triton into orbit around Neptune.

In all this discussion, Pluto remains as a mere afterthought: it is an even smaller, colder body of the type of Titan or Triton, with only a feeble trace of atmosphere and very low internal temperatures. Pluto has a large satellite, Charon, about half Pluto's diameter, that is rotationally locked into a 1:1:1 spin-spin-orbit resonance: both Charon and Pluto always keep the same face toward each other as they orbit around their common center of mass. The Pluto day, the Charon day, and the Charon month are all exactly the same length.

IF THERE is life in the sub-ice oceans of large satellites, how might we know of its existence? Europa's thin, mobile crust shows vast numbers of cracks and grooves that have been filled by water that has welled up into crevices and frozen there. Most, if not all, of Europa's crust has been cycled in this manner. Whatever dissolves, floats, swims, or dies in that water could be delivered to the surface in a quick-frozen state by natural processes. But we have not yet attempted to land a spacecraft on the surface of Europa. Because of Europa's proximity to Jupiter, a spacecraft arriving from Earth would pick up a substantial excess velocity while falling in to Europa's orbit. A would-be lander would somehow have to dissipate that extra energy, either by the judicious firing of rocket engines or by repeated, carefully choreographed passages by the Galilean satellites, in order to match orbits with Europa. Then a further major engine burn would be required to land at low, survivable speeds. All the time the spacecraft is maneuvering in the inner Jovian system, it is accumulating radiation damage from Jupiter's intense radiation belts. This is not an altogether impossible task, but it is very ambitious.

How then might we know if there is any life there? Freeman Dyson of Princeton's Institute for Advanced Studies, one of this century's authentic geniuses, has suggested a little less than half jokingly that impacts on Europa could from time to time eject freeze-dried alien "fish" into orbit around Jupiter; perhaps, under the long-term gravitational tweaking of the Galilean satellites, into orbit around the Sun. For our purposes, we should interpret "fish" as any distinctively biological organic matter or tissue, not necessarily an intact specimen large enough to be served for dinner. (It would not be wise to eat an alien fish in any case: even if it is biochemically similar to terrestrial life, and even if terrestrial and Europan life derived from the same origins, alien products of billions of years of independent evolution would contain amino acids, sugars, and organic bases that terrestrial life forms cannot metabolize, and lack others that are essential to our diet. A wide variety of toxic materials may also be present in their flesh. Would you buy billion-year-old freeze-dried alien fugu sashimi from a complete stranger?) Dyson's fish would be considerably more accessible to terrestrial spacecraft in their orbit around Jupiter than on the surface of Europa: the problem is that we haven't a clue where to look for it. To paraphrase something John Wayne might have said, "There's a lot of space out there for a bad fish to get lost in."

Suppose a piece of this alien fish-stuff were to drop out of the sky, survive entry into the atmosphere, and land, all steaming and succulent, on Earth. I can see it now on the menu at the Explorers' Club: Blackened

Reconstituted Dried Irradiated Europan Sole. Is this what Hamlet meant when he wrinkled up his nose and complained that "something is rotten in the state of Denmark?" Probably not. But if such an injection occurred, would we notice, and could we deduce what the alien matter was?

Since I can't answer that question, and since my readers are as yet in no position to object to this sleight of hand, I will hasten to change that question into something I *can* answer: "Has anyone, irrespective of their credibility, ever reported the fall of organic matter from the sky?" Since more or less everything that can be formulated in language, however improbable or impossible, has in fact already been reported by someone, the answer to this question is a hearty yes. Such questions are the peculiar domain of Charles Fort, a man who made a career out of collecting tidbits from newspapers and magazines, selected to be embarrassing (unexplainable) to establishment science. Whether you take him seriously or not, his books (*The Book of the Damned, Lo, Strange Talents,* and *New Lands*) make fascinating, often hilarious, reading. His eclectic research provided him with colorful snippets of reports of storms of frogs and fish, rain of blood, black snow, falls of globs of gelatinous goo (the G3 phenomenon), and so on. But strangest of all are reports of falls of chunks of meat and flesh, much of it in an advanced state of decay. Fort's work, incidentally, is not unique in human history: he is in many ways a modern counterpart of Pliny the Elder, whose lively and episodically uncritical accounts of similar strange events ("fiery javelins" in the sky; falls of "wool") in his *Historia Naturalis* fit comfortably on the same bookshelf with Fort's works. Fort packages these reports in a fictional, tongue-in-cheek, highly bizarre context of galactic warfare, with casualties from such conflict falling to Earth. But that is just part of his game of tweaking the noses of the scientific community. He is always toying with his readers, winking and smirking at us as he spins his fantasies.

If it were not for the fact that many of the phenomena cataloged by Fort (falls of frogs and fish, black snow and rain, falls of "blood") have now entered the domain of modern science (as the products of waterspouts, industrial pollution, and rain stained by red desert dust), we would be justified in ignoring these bizarre reports. Even Fort's reports of the G3 phenomenon have some tangency to reality: these falls tend to occur at times of major displays of cometary meteor showers. They were reported by professional scientists in such serious journals as the *American Journal of Science*. A century and a half after those reports, the mass spectrometer carried to Halley's Comet in 1986 by the European Space Agency's Giotto mission found enormously long chains of formaldehyde molecules, the vapor of a polyformaldehyde material that looks, in the technical language of laboratory physical chemistry, "exactly like G3."

So let us suspend our disbelief for a moment and consider whether it is ever possible for "flesh" to fall from the sky. The answer, of course, is yes. There are credible eyewitness reports of birds in flight being killed by meteorites; frogs and fish have been unwitting passengers in waterspouts; tornadoes have been seen to pick up animals as large as a cow and deposit them more or less catastrophically back on the ground. Although I have not yet heard of a tornado picking up and dismembering a dead cow, I am perfectly willing to believe that it has happened. In other words, I am willing to believe that various forms of flesh, in diverse states of preservation, have in fact been observed to fall from the sky by sane and sober observers. But are these gobbets ever visitors from outer space? I would require some form of proof more biologically profound than the opinion of the farmer who found the stuff in his fields, or the ever-tongue-in-cheek Fort. I would listen carefully to expert testimony on the chemical, cellular, and genetic properties of such material—but no such evidence yet exists.

HAVING LOOKED over the range and variety of large satellites in the Solar System, we are acutely aware that there are many plausible variations on these bodies that may be found elsewhere. For example, bodies may exist having compositions within the range observed in our Solar System, but not the same as any known body. Similarly, they may lie modestly outside the observed composition range without violating our understanding of chemistry and physics. They may differ in composition by reason of different condensation temperatures and different conditions of accretion. They may even be identical in composition to known Solar System bodies, but differ in size: as we saw in our comparison of Earthlet, Earth, and Earthissimo in Chapter 8, the structure, present composition, and evolutionary histories of these bodies are remarkably dissimilar despite their identical initial compositions. Finally, bodies that started out closely similar may diverge because of different present environmental conditions, such as temperature, distance from the central star, tidal interactions, and bombardment history.

Three major classes of such variants on icy satellites seem especially attractive as potential abodes of life. First, large nitrogen- and methane-bearing satellites ("Titanoids") located a little closer to the Sun than Titan could easily have much more massive atmospheres and a much stronger greenhouse effect, possibly even to the extent of having liquid water-ammonia oceans on their surfaces. Second, satellites with sub-ice oceans ("Europoids") might exhibit a remarkable degree of autonomy by being very insensitive to the external conditions imposed by their sun. Third, bodies of grossly Earthlike composition ("Earthoids") in orbit around brown dwarfs may be maintained at congenial temperatures by their primaries. The latter scenario, however, is complicated by tidal interactions and by the slow cooling of the brown dwarf.

It is time we looked more closely at what kinds of orbits are most suitable to producing stable, life-friendly conditions.

12

DIPPY AND MERELY ECCENTRIC MOTIONS

THROUGH THE FIRST FEW CHAPTERS, we have emphasized the vast variety of materials out of which worlds can be made and the importance of planetary size. Our focus has been on the intrinsic nature of planets, neglecting their nurture: we have had relatively little to say about where these planets reside and how they move, beyond stating the most obvious fact that the temperature of a planet is strongly influenced by its distance from its star. Now we will inquire into what kinds of orbits, and what kinds of planetary dynamics, are conducive to stable, moderate, life-supporting conditions on the surfaces of planets.

THE SIMPLEST possible orbit of a planet around a star, or of a moon around a planet, is a circle lying in the plane of the larger body's equator and centered on that body. As we discussed in Chapter 2, many of the moons, and some of the planets in our Solar System, follow orbits that are nearly circular. Also, most bodies in the Solar System and in the satellite families within it orbit close to the equatorial plane of their primary body (the body around which it orbits). This plane in our Solar System, which for convenience is taken to be the plane of Earth's orbit, is called the ecliptic. Since a moon or planet in circular orbit is at a constant distance from its primary, it neither rises nor falls in the primary's gravity field, and therefore never gains or loses energy of motion: it orbits at a constant speed and at a constant angular rate. Earth, for example, completes a trip around the Sun, 360° of arc, in 365.24 days, just a hair less than one degree of motion along its orbit per day. Earth's orbit is close to circular, but not perfectly so. Its mean distance from the Sun is defined as 1 AU, which is about 150 million kilometers, or 93 million miles. Earth actually follows a slightly elliptical orbit that takes it out to a maximum distance of 1.015 AU and in to a minimum distance of 0.985 AU from the Sun. These small excursions give Earth's orbit an eccentricity of 0.015. Among the planets, Mercury's eccentricity of 0.2056 and Pluto's of 0.2566, at the inner and outer fringes of the planetary system, are by far the highest. Of the other planets, only Mars, with an eccentricity of 0.09, exceeds 0.06. The eccentricities of these three planets are large enough to cause the intensity of sunlight falling on them to vary seasonally. The most important single effect of eccentric orbits is the annual cycle of solar heating. Low eccentricities favor constant heating and a stable climate.

Mars has an orbital eccentricity of 0.09. As Mars orbits the Sun near its average distance of 1.52 AU, the intensity of sunlight drops off with the square of its distance from the Sun. That distance ranges from a minimum of 1.38 AU at perihelion to 1.67 AU at aphelion. The average intensity of sunlight on Mars is 590 watts per square meter, much lower than the 1370 watts per square meter falling on Earth. But, because of the eccentricity of Mars's orbit, the intensity of sunlight varies greatly, from 718 W/m^2 at perihelion to only 494 W/m^2 at aphelion. The entire planet goes through synchronized perihelion heating and aphelion cooling, with both hemispheres warming and cooling in phase with each other.

Very high eccentricities, however, may sometimes permit planets or large moons (let us call them collectively "worlds") to cross the orbits of their neighbors. If our Solar System is representative, an eccentricity of about 0.25 is sufficient to get most planets into trouble with their neighbors. In the Solar System, trespassing on another world's turf is a serious matter, since it makes disturbing close encounters and even violent collisions possible. High eccentricity has less dramatic consequences in planetary systems in which planets are few and far between. But climatological consequences are still there.

There is a special case in which eccentricity is not lethal: when the orbital period is quite short. When there are only a few Earth days per year, an atmosphere of moderate opacity and heat capacity can smooth out the temperature variations.

IN ADDITION to being eccentric, orbits may also be inclined relative to the equators of their primaries. We know, from the study of asteroids, that small bodies sometimes have orbits that are tilted far outside the plane defined by the equator of the central star or the general plane of motion of the star's planetary system. Are such orbits a natural result of planetary-system origin and evolution? Are highly inclined orbits good or bad for a planet with pretensions of habitability? How does the neighborhood influence property values?

Inclination by itself has no significant effect on the temperatures and climate of the orbiting world; however, in a system with large amounts of debris orbiting in the equatorial plane (a dust disk around a young star, a ring system around a planet, or an asteroid belt around a mature star), a world in a highly inclined orbit may experience serious hazards. Whereas a low-inclination world may spend all its time embedded in the debris disk, it is moving along with the small bodies. Its velocity relative to these small bodies is then very low. But, if the world's orbit is highly inclined, it will briefly traverse the debris disk twice per orbit at high speed. Encounters will be far less common and far more violent.

The baseline planetary system, formed from an extensive flattened nebular disk, naturally assembles planets in orbits close to the central plane. Only bodies far out on the fringes of the system can resist the leveling influence of the disk. In our own system, many of these bodies may have been ejected from the disk into extremely elongated orbits by close gravitational interactions with the half-grown gas-giant planets. Once ejected into orbits with periods of hundreds of thousands of years, these cometary bodies arc outward from the Sun, loiter near the aphelia of their trajectories, and then reluctantly fall back into the Solar System, only to find that the Solar Nebula has evolved and expired before that first great orbit was complete. Arriving late for the banquet of planetary growth, these comets return after all the food is gone and fail to grow further. Any of these bodies may have a high orbital inclination, but none should have planetary size, benign surface temperatures, or stable temperatures. The interiors of the very largest of such bodies (many hundreds of kilometers?) may begin to melt early in Solar System history, but they then cool and freeze completely at an early date. As such, they are biologically uninteresting, except insofar as they carry with them large amounts of ancient, nonbiological organic matter.

Leftover small solid bodies formed at even greater distances from the Sun become long-period comets, many with orbital periods of several million years. These comets have random orbits around the Sun, about half traveling in retrograde orbits, opposite in direction to the sense of orbital motion of the planets. Clearly they represent a significant collisional hazard to all the planets.

Gravitational perturbations of small bodies in near-circular, low-inclination orbits (asteroids, for example) can also occasionally generate highly inclined orbits. Although the vast majority of asteroids have orbits inclined less than 20° to the plane of the Solar System, there are a few that have been sent on thrill rides by close planetary flybys or orbital resonances. A few asteroids, including Euphrosyne and Aethra, have inclinations that cluster around 26°. Very few are more inclined: Olympia, Mireille, Barcelona, Zerlina, Lick, Midas, Sisyphus, Hidalgo (which orbits close to Jupiter), and Troilus (which follows Jupiter's orbit) at inclinations of 30° to 43°. The belt asteroid Betulia orbits at an inclination of 52.9°, and the Apollo (Earth-crossing) asteroid 1975 YA at an astonishing 64.0°. Looking at the entire known population of asteroids, the odds against one having an inclination greater than 33° are a thousand to one.

In the Solar System, only small bodies have achieved such high-inclination orbits. The asteroids in the belt were, very roughly speaking, discovered in descending order of brightness, for obvious reasons: low catalog numbers tend to indicate bright (and usually large) asteroids. Only one of the one hundred first-discovered asteroids has an inclination as high as 26°. Bodies of planetary size, which accrete from astronomical numbers of smaller bodies, have orbital properties that tend to average out the orbital excesses of the small bodies out of which they formed. As with humans, maturation usually has a leveling effect on eccentricity. For a full-grown planet to be perturbed into a high-inclination orbit it would have to survive a very close encounter with Jupiter or Saturn. That planet would then continue to cross Jupiter's orbit, risking a collision or another strong kick. Every such encounter would have a catastrophic effect on the planet's orbit and hence on its climate. The orbit could not be expected to remain stable for very long. This is a high-risk business indeed.

There are certain other hazards associated with high-inclination orbits. For example, we mentioned that a highly inclined body would pass through the plane of the system at very high speeds. A collision with an asteroid with a thousand times smaller mass would be so violent that the larger body would be destroyed, reduced to rubble that could be cleared away by solar radiation pressure.

Among the planets of the Solar System, the most highly inclined is Pluto, at 17.1°. But Pluto has other, more severe problems than just inclination: its orbit is so eccentric that it actually crosses the orbit of Neptune. From about 1979 to 1999, Neptune was the farthest planet from the Sun. Because of a resonant (harmonic) relationship between the orbits of Neptune and Pluto, these planets actually never get very close to each other, but follow a choreography that repeats over long periods of time. Here at last is a concrete example of the harmony of the spheres.

There is another, more subtle effect of inclination: a planet in a highly inclined orbit would spend much less time in the high-speed equatorial solar wind, and much more time at high solar latitudes. This could somewhat reduce the total solar wind exposure of the planet, but we know of no circumstances in which this could significantly affect the climate stability or habitability of a world. Compared with the collision hazard, this is a minor effect.

In systems much simpler than ours, with only one or two massive planets, there may be no well-defined orbital plane, or that plane may not coincide with the equatorial plane of the star. In such a case, there is no particular significance to high inclination.

QUITE ASIDE from the properties of the orbits of worlds, the conditions on their surfaces are strongly affected by the way in which the planet rotates. Both the length of the day and the inclination of the spin axis are important. For example, Mars, like Earth, has a pronounced tilt of its spin axis: the plane of the planetary equator is not the same as the plane of its orbit. Axial tilt causes one pole to point generally toward the Sun for half an orbit, then generally away from the Sun for the other half of the orbit; in other words, it causes the planet to experience seasonal cycles. The hemisphere tilted toward the Sun will experience summer at the same time that the opposite hemisphere has winter. The axial tilt of Mars, 25° compared to Earth's 23.5°, causes seasonal effects similar to those experienced on Earth. Some planets may have much more extreme axial tilts and much more extreme seasonal cycles.

Extreme axial tilt by itself causes temperature problems. The most extreme example in our Solar System is Uranus, with an inclination of 98°. That axial tilt places the pole within 8° of the orbital plane: as Uranus executes its eighty-four-year orbit, its poles take turns pointing almost directly at the Sun. For forty-two Earth years at a time the Sun never sets at the north pole of Uranus, and never rises at the south pole. Then the equatorial plane of Uranus sweeps across the Sun, and the reverse situation persists for another forty-two years. The overall result of this strange geometry is that the poles of Uranus receive more heat from the Sun over the course of a Uranian year than the equator does. The poles are the warmest places on the planet, and the equator is the coldest! On Earth (and Venus, Mars, Jupiter, Saturn, and Neptune) the driving force behind the circulation of the atmosphere is the transport of excess heat from the equator to the poles. But on Uranus, the motions of the atmosphere serve the exact opposite function.

On a terrestrial planet, where internal heat is not important and the mass and opacity of the atmosphere are not too large, high axial tilts can cause enormous temperature extremes. The fraction of the planetary surface that can maintain moderate conditions diminishes rapidly for axial tilts over about forty degrees.

Thus there are two distinct kinds of seasonal variations at work on Mars. In addition to the Earthlike seasonal effects of alternating summer and winter in each hemisphere, caused by the inclination of the spin axis, there is the global warming and cooling cycle caused by Mars's varying distance from the Sun. Clearly, if one hemisphere is tilted away from the Sun at aphelion, it will experience both "winters" at once. Suppose that the north pole is tilted directly toward the Sun at the time of perihelion passage. The intensity of solar heating of all of Mars (due to perihelion passage) and the intensity of solar heating of the northern hemisphere (due to it being northern midsummer) will reinforce to provide exceptionally warm northern hemisphere summer temperatures. At the same time, the midwinter temperatures at the south pole will be moderated by the fact that the entire planet is getting much more heat from the Sun. Half a Mars-year (278 Earth days) later, the north pole is tilted away from the Sun and the entire planet is colder because Mars is now at aphelion. The northern hemisphere winter will be exceptionally severe, but the southern summer will be moderated by the compensatory effects of summer exposure but diminished intensity of sunlight.

As if this were not complex enough, the direction of any planet's spin axis changes slowly with time. Like a toy gyroscope spinning in Earth's gravity, the spin axis of a planet describes a circle on the heavens, centered on the pole of the planet's orbit. This motion, characteristic of spinning bodies that feel external forces, is called precession. Basically, under the influence of the Sun's gravity, Mars's spin axis precesses slowly around the pole of its orbit once every 50,000 years. This also is closely similar to Earth, where the precession cycle lasts 26,000 years. Consider two different states that can arise during the 50,000-year precession cycle. In the orientation described above, with the north pole tipped toward the Sun at perihelion, there will be an era lasting thousands of years in which the northern hemisphere of Mars will suffer exceptionally severe temperature swings, while the southern hemisphere will have exceptionally moderate temperature cycles. But half a precession cycle (about 25,000 Earth years, or 13,300 Mars years) later, the geometry will be reversed: the north pole will now be tipped away from the Sun at perihelion, and the south pole will be tipped toward the Sun. The southern hemisphere will then suffer wild temperature excursions, while the northern hemisphere will have moderate summers and winters.

A second complexity is due to the fact that the orbital inclination and eccentricity of a planet, and even the inclination of its orbit, can change with time. The polar tilt of Mars can range from about 15° to 35°, and the eccentricity of its orbit may range from zero to about 0.12. Indeed, in some theories the axial tilt can on occasion be as large as 45°. At an axial tilt of 45°, the average annual temperature at the poles is essentially the same as at the equator, but the annual range of temperatures is profound. The very small range of axial tilt variations of Earth (ranging from about 22.5° to 24.5°) is directly attributable to the effects of the Moon. In other words, without the Moon, Earth's axial tilt would vary as wildly as that of Mars, and Earth's climate would be far less stable than it is. Having a large Moon makes Earth both more habitable and more romantic.

THE PHENOMENA that we have introduced in referring to the example of Mars are of very broad relevance. Eccentricity causes climate problems by varying the intensity of sunlight. Those variations can be moderated if the heat capacity of the planet's atmosphere is so large that temperatures cannot change much over a year. This is very definitely not true of many locations on Earth. Desert regions, where there is little water vapor to provide an infrared blanket above the ground, often cool by radiation at an impressive rate, sometimes dropping by more than forty Fahrenheit degrees (twenty-two Celsius degrees) overnight. Mars, with its very tenuous atmosphere, has much wider day-night temperature cycles than any place on Earth.

Wide temperature extremes can lead to the sequestration of water or other condensable volatiles in cold traps such as polar ice deposits. On a water-rich planet, cooling one region to exceptionally low temperatures will precipitate out vast quantities of snow, drying and cooling the entire atmosphere of the planet. An ice age will begin in the cooled regions, the reflectivity of the planet will be increased by the highly reflective ice surface, the greenhouse effect will be diminished by the lowered water vapor content of the atmosphere, and a climate crash will ensue. To avoid cooling by a catastrophic amount, say one hundred Fahrenheit degrees over a one-hundred-day season, cooling rates less than a degree per day are required. Such low cooling rates require immensely massive, opaque atmospheres.

Interestingly, if Earth had as high an orbital eccentricity as Mars, the surface climate would be so unstable that life would be severely curtailed—but there would still be some places on Earth that would maintain stable conditions. The best place to be on Earth would be at the hydrothermal vents in the ocean floor. Even in the extreme case that the entire top kilometer of Earth were completely sterilized or frozen by an era of wild temperature excursions, bacteria in groundwater and advanced life forms on the ocean floor could eventually repopulate the planet. In other words, Europas are highly tolerant of orbital eccentricity.

We mentioned the possibility that having a short orbital period helps average out the temperature excursions caused by high axial tilt. But the situation is both more complex and more promising than this simple statement suggests. Consider the example of Mercury. We expect that the fringe planets in a system will, on the average, have relatively high orbital eccentricity. In other words, short orbital periods, which imply unusual proximity to a sun, will often be rather eccentric. Mercury shows us the logical result of eccentricity: the strength of the tidal forces exerted by the central star varies greatly over the orbit, being by far the strongest at perihelion. Since eccentricity inevitably causes the angular speed of the planet around its orbit to vary (a direct result of the principle of conservation of angular momentum), there is an opportunity for the planet to adjust its spin so as to keep the planet's tidal bulge aimed at the star around the time of perihelion passage. Such a lock means that the (constant) angular rate of the planet's spin is not the same as the average angular rate of motion around its orbit, but in fact is closely equal to the *maximum* orbital angular rate; i.e., that near perihelion. For such a lock to occur, the rotation must be prograde (in the same sense as the orbital motion) and faster than the average rate of orbital motion. One or the other end of the tidally distorted planet must point at the Sun at every perihelion passage. In other words, there must be a spin-orbit resonance, and, for a significantly eccentric planetary orbit, the resonance cannot be 1:1. Since the spin rate (the first number in the ratio) must be greater than the mean orbital rate (the second number), resonances such as 1:2, 2:3, 3:4, 2:5, and so on are ruled out. Among the possible resonances are 3:2, 2:1, 5:2, 3:1, 7:2, and so on. Resonances such as 4:3, 5:3, 7:3, 6:5, 7:5, and so on are also not allowed, because the planetary tidal bulge would not point at the Sun at every perihelion passage. A more transparent way of showing the connection between spin rate and orbital period is to write the plausible resonances as 3:2, 4:2, 5:2, 6:2, 7:2, and so on. This method clearly shows that there must be an integral number of half-spins per orbit (i. e., 3/2:1, 4/2:1, 5/2:1, 6/2:1, 7/2:1, and higher resonances of the same form).

The higher the spin rate (the greater the angular speed of rotation), the greater the orbital angular speed of the planet must be at perihelion for a lock to occur. The higher the resonance, the more eccentric the orbit of the planet must be. A perfect 1:1 resonance implies an eccentricity of zero (a perfectly circular orbit). A 3:2 resonance in which the angular rotational speed at the moment of perihelion passage exactly matches the angular orbital speed would require an eccentricity of 0.225. But it is not only the instant of perihelion passage that matters. The best match is the one in which the average fit to perfect lock for the time bracketing perihelion passage is the best: for Mercury, the eccentricity of 0.20563 assures a small mismatch of speeds precisely at perihelion, but a better *average* fit across the perihelion region. An observer riding on the tidal bulge would see the Sun rise rapidly above the horizon and slow as it approaches the zenith, overshoot the zenith by a few degrees, stop and back up to the other side of the zenith, passing the exact zenith at perihelion (noon) going in the backward (retrograde) direction, then accelerate back across the zenith a third time and dive below the horizon opposite where it rose. The *average* distance of the Sun from the zenith is smaller this way than if the angular rates were perfectly matched at perihelion and no retrograding took place.

The other allowable resonances require higher eccentricities: 0.412 for perfect angular-speed fit to the 4:2 resonance, 0.581 for 5:2, 0.732 for 6:2,

0.871 for 7:2, and so on. The optimum eccentricity would in each case be a little less than these numbers, just as it is for Mercury. Clearly the higher resonances, which are otherwise dynamically sensible, can cause major orbit-crossing problems in a multiplanet system. Mercury, at its present orbital period, would reach out to Venus at aphelion if its eccentricity were 0.868. When you're a planet, it isn't nice to reach out and touch someone. Even worse, with so high an eccentricity, Mercury would drop in to 0.0511 AU from the Sun at perihelion, enough to raise the noontime equatorial temperature to 2000 K. This is hot enough to melt any rock type, including native steel alloys, and to generate a significant atmosphere of rock vapor. Of course, this problem would be reduced if the central star were a lot cooler than the Sun — but then a planet in so eccentric an orbit would freeze at aphelion!

The cure for ailing orbital elements or faulty spin is not easy. Planetary spin-doctoring is affordable for small asteroids but not for full-sized planets. Likewise, changing a world's orbital inclination or eccentricity is prohibitively expensive. It is far better to invest in a good planet in a nice neighborhood than to buy a handyman's special. And choosing a good neighborhood in astronomy means knowing a lot about your local star.

13

'TIS THE STAR, THE STAR ABOVE US, GOVERNS OUR CONDITION

WE ARE SO ACCUSTOMED TO our yellow, constant Sun that we think it is representative of suns in general. It isn't. Since Giordano Bruno popularized the notion that the stars are Suns similar to our own, many have wondered about their nature and variety. The demographics of the star population of the spiral arms of the Milky Way, which are fairly well known, tell us that the Sun is a member of a minority group. Since most stars are very different from the Sun, we must consider Shakespeare's pronouncement, given as the title of this chapter, and inquire how other types of stars might govern the conditions on their planets.

Most stars, like our Sun, belong to the main sequence (MS). They are reasonably stable hydrogen-burning stars that derive their energy from the fusion of hydrogen to helium by either (or both) of two different mechanisms, the proton-proton chain (most important in stars with masses less than about 1.5 Suns) and the carbon-nitrogen-oxygen catalytic cycle (most effective above 1.5 Suns). These stars form with a wide variety of masses, from the borderline for nuclear reactions around 0.01 solar masses through the feeble beginnings of hydrogen fusion at 0.08 solar masses on up to enormously luminous stars with sixty or more times the mass of the Sun. Some 90 percent of all stars belong to the main sequence.

The main sequence stars are subdivided according to temperature or color into seven principal classes: O, B, A, F, G, K, and M. Each successive class is cooler, less massive, smaller, fainter, and longer-lived than its predecessor. The hottest and most massive, the O stars, have masses of

20 to 100 Suns and luminosities up to several million Suns. They are so hot that most of the light they emit is ultraviolet radiation: they appear violet-white to the eye. The O stars, like all the other spectral classes, are subdivided into finer categories labeled 0 to 9: the very faintest, smallest, coolest O stars belong to class O9. An O5 star has about sixty solar masses.

The next class of main sequence stars in descending order of temperature, luminosity, and mass are the blue-white B stars, with average masses of about 6 Suns (in a range of about 3 to 20) and luminosities of about 1000 (in a range of 50 to 40,000) Suns. Then come the white A stars, 1.7 to 3 times the mass of the Sun, with luminosities of 10 to about 50 Suns. Next are the white but somewhat cooler F stars, with masses of about 1.1 to 1.7 Suns and luminosities of about 1.4 to 7 Suns.

Next come the yellow-white G stars, one of which is our Sun, a G2 star. The G stars have masses of 0.8 to 1.1 Suns and luminosities of 0.5 to 1.4 Suns. The next cooler MS stars belong to the orange K class, with masses of about 0.4 to 0.8 Suns and luminosities of 0.04 to 0.5 Suns. Finally, the very coolest, red MS stars, belong to the M class. The M stars, which are very abundant, range from 0.08 to 0.4 times the mass of the Sun and have luminosities of 0.00001 to 0.04 Suns. The very faintest of all stars, M9 dwarfs, are almost impossible to detect over normal interstellar distances.

There are also several other systematic differences between these classes of MS stars: the O, B, A, and some F stars tend to be rapid rotators, whereas G, K, and M stars tend to rotate very slowly. Also, all except the very largest MS stars are detectably variable in brightness. Our Sun varies by less than 1 percent as a result of sunspot and solar flare activity, but variability becomes much more important for lower-mass stars. M5 dwarfs may brighten by a factor of two or three during flare eruptions. Their surfaces are largely covered by sunspots much of the time. Every MS star of interest to us as a potential home for life is a variable, but the M stars are by far the most variable.

The main sequence star types are by no means equally common. Of every million MS stars, about a hundred are members of the O class (and therefore visible over enormous distances). The same sample contains a few hundred B stars, perhaps 2500 A stars, 10,000 F stars, 40,000 G stars, and 150,000 K stars. All the rest (about 800,000) are faint red M stars. Our Sun, a G2 star, is in the ninety-sixth percentile in mass and luminosity, far brighter than the typical star.

If main sequence stars make up nearly 90 percent of all the stars known, then what are the others? Almost all the others are white dwarfs, stars that have exhausted their hydrogen fuel and fallen off the main sequence. Also off the main sequence are the highly visible giant and supergiant stars such as Betelgeuse, Aldebaran, and Antares. Such stars, despite their visual prominence, are actually extremely rare in the Galaxy, making up far less than 1 percent of the population.

Pick a random star and you will very likely get a cool, faint M star, difficult to see even if it is only a few parsecs away. But pick a random star from those you see in the night sky and you will probably choose an extremely rare, highly luminous star like Canopus, one hundred million times as bright as that nearby red dwarf but ten thousand times as far away.

Clearly the large majority of the stars in the galaxy are M-class main sequence stars. What can we say about planetary systems in orbit around such stars?

The most extreme red MS star, of class M9, has a mass of about 0.07 Suns and a luminosity of about 0.00001 Suns. The star as seen from its planets would be a very dull, muddy red, with great irregular dark blotches covering most of its face. From time to time, bright, hot flares would erupt from the star and increase its brightness by a factor or two or three. A planet at 1 AU from an M9 hydrogen-burning star would receive about as much energy from its star as Earth receives from a full Moon, but most of that would be a weak trickle of infrared radiation, invisible to the eye. The temperature of Earth would drop to about 18 K. Water vapor would condense and fall as snow, and carbon dioxide would condense and deposit a layer of dry ice powder on top of the snow. Then nitrogen and oxygen would rain out and freeze, coating the surface of Earth with a frosting of solid nitrogen and oxygen several meters thick. Argon would condense and also fall as a sparse snow. The only atmosphere left would be a wispy trace of hydrogen, helium, and neon.

The only way to maintain reasonable temperatures on the surface of a planet around an M9 star is to place that planet very close to the star. For example, suppose we want to have a surface temperature similar to present conditions on Earth. We would need to have an Earthlike planet at a distance of 0.003 AU from the center of the star. That is only 450,000 km, about the distance of the Moon from Earth. The star itself is smaller than Jupiter or Saturn, only about 50,000 km in radius. The star as seen from the planet spans nearly thirteen degrees. The apparent area covered by the disk of the star would be 530 times as large as the apparent area of the Sun as seen from Earth. The presence of 0.07 solar masses in a sphere with a radius of 50,000 km gives the star a density of 330 grams per cubic centimeter, about 30 times as dense as lead. The tidal forces exerted by this compact, close star are immense, easily sufficient to despin the planet quickly.

The enormous surface gravity of the star, 430 Earth gravities, crushes its atmosphere into a thin layer in which the pressure rises rapidly with depth. The escape velocity from the surface of the star is over 650 kilometers per second. Very high speeds are needed to orbit close to the surface of so dense a star. The orbital speed of the planet, 154 kilometers per second (five times Earth's orbital speed), takes it all the way around the star in five hours. The day and the year are both five hours long!

The temperature of the star, about 2300 K, is so low that molecules such as CO, nitrogen, OH, water vapor, and metal oxides including AlO, TiO, and the like must be abundant. Almost all of these molecules have strong absorption bands in the red or infrared part of the spectrum. Many minor constituents of its atmosphere may condense to form wispy clouds of refractory metal and oxide minerals. Think of it: a cloudy star!

But we have barely tapped the strangeness of this system. There is a slightly more subtle but absolutely fatal flaw in this carefully engineered system: the tidal forces of the star are so gigantic that even at the surface of the planet the gravitational attraction of the star surpasses that of the planet by more than a factor of five. An object or person on the surface of the planet would be snatched from the planet by the star's intense gravity. In fact, the planet's atmosphere, oceans, mountains, crust — everything! — would be whipped away. The problem is that, in order to have livable surface temperatures, we have had to place the planet so close that the gravity of the star tears the planet apart. This phenomenon of tidal disruption, which was first explored mathematically by the French mathematician Edouard A. Roche in 1850, explains why the rings of Saturn do not accumulate into another large satellite. In the present application, it tells us that an Earthlike planet in the comfort zone around an M9 main sequence star would be instantly disassembled into a belt of rubble that would dwarf the mass of our asteroid belt by several thousand times. But this instant asteroid belt would be crammed into a volume of space about ten thousand times smaller than that occupied by our belt.

The problem of tidal stripping persists for all other M-class stars up through class M7 (about 0.0001 Suns luminosity, and about 0.14 Suns mass), but tidal despinning continues to be important even beyond that point. In addition, the larger M stars could accommodate several types of spin-orbit resonances.

Stars between spectral class M7 and F, including all the G and K stars, are plausible hosts for planets with Earthlike temperatures: Earths placed at the correct distance to have normal terrestrial temperatures would not be torn apart by tidal forces. The F stars of most interest to us belong to classes F7, F8, and F9, which have masses ranging up to about 1.3 Suns and luminosities up to about 2.5 Suns. The very largest F stars (F0), with masses of about 1.7 Suns, have luminosities of about 7 Suns. Although the largest F stars are plausible hosts for habitable planets, we should keep in mind that they are relatively rare: stars between classes F0 and F6 make up only about 0.4 percent of all MS stars. The question of whether such systems are habitable becomes less cogent when we realize they are not only rare, but also have relatively short lifetimes. The more massive MS stars burn their hydrogen fuel at so furious a rate that they do not live very long.

We are pretty well satisfied that our G2-class Sun can support habitable planets for extended periods of time. At least, we hope we are right in this judgment. But what about main sequence stars more massive and luminous than the Sun? The most extreme MS stars, with luminosities of millions of Suns, are of little interest because there is almost no time of stable residence on the main sequence. For example, an O0 ("oh zero") star with one hundred times the mass of the Sun would have a luminosity of about ten million Suns. Since it contains one hundred times as much hydrogen fuel as the Sun and burns it at a rate ten million times as great, it will exhaust its fuel supply in only one hundred thousand years, a time comparable to that required to assemble the star in the first place, too short for the accretion of large terrestrial planets, and far too short for the evolution of complex life forms. After departing from the main sequence, the star will briefly dabble in supergiant exhibitionism, then detonate as a supernova explosion. Incredibly, there is a further problem with so massive a star: the fierce radiation pressure in its interior, generated by the incredible rate of hydrogen fusion, can literally blow the star apart. It is quite impossible to imagine advanced life forms originating in such a system.

The dilemma is this: for a star to have biological interest, it must remain stable on the main sequence long enough for life to arise and evolve to complexity. If we knew how long this process takes, we could specify the maximum-mass (minimum-lifetime) star capable of supporting life-bearing planets. We can crudely narrow down the search by considering a few facts and opinions gleaned from the study of our own Solar System.

First, how long did it take the Sun to become a stable main sequence star? Even after ten million years, the T-Tauri-phase activity of the early Sun was still significant. By one hundred million years, the Sun should have been stable enough to be a responsible host to planets and life.

How long did it take to form the planets? Most estimates based on mathematical models of planetary growth have suggested about one hundred million years, although some give answers as short as a few million years. But those starting conditions that give the most rapid accretion are also those that produce an Earth whose surface is molten, far too hot for interesting chemistry. Cooling times of tens of millions of years seem reasonable. Core formation occurred within a few tens of millions of years, suggesting that there was no stable crust before that time. The giant collision that made the Moon unfortunately cannot be dated with useful accuracy, although we know it was at least 4 billion years ago. But in any case that particular constraint may be idiosyncratic to Earth, and not directly applicable to other worlds. The earliest known rocks on Earth date back to 4 billion years ago (600 million years after the origin of the Solar System). The earliest evidence for life on Earth dates back to the same rocks, whose chemistry suggests that the ocean and atmosphere were already well established. The earliest known fossil bacteria date back about 3.5 billion years. By 3 billion years ago, photosynthetic algal colonies in the tidal margins of the oceans began to build large calcareous structures called stromatolites. Recognizable animals appear only in the upper Precambrian, less than a billion years ago. Therefore if we draw the line between "biologically boring" and "biologically interesting" at the point of emergence of bacteria, it took about a billion years for Earth to get that far.

It also took a considerable time for the impact rate on Earth to drop low enough to permit the stable existence of life. Recent theoretical calculations on the early history of Earth suggest that large impacts (bodies tens of kilometers in diameter, not the size of the Moon or Mars!) may have heated the biosphere to sterilizing temperatures several times in the first billion years. Perhaps we should speak not of the origin of life on

Earth, but of the repeated replenishment of terrestrial life after multiple global catastrophes. In that case, we might conclude that it really only takes one hundred million years or so for life to advance to the point of making bacteria, and that a planet spared the intense early bombardment history experienced by our inner planets might have accomplished truly impressive evolutionary advances in a few hundred million to a billion years. Alternatively, we might conclude that the tail end of the assembly process of Earth-size planets just naturally takes a billion years to play out. The overall result of this discussion is to confess we really aren't sure how long it takes to make interesting life forms (even if, miraculously, we should be in perfect agreement as to what level of complexity is interesting), but a variety of arguments suggest time intervals of one hundred million to a billion years. This is the minimum stellar lifetime that we need.

Now we can return to the main sequence. What is the largest star that can reside on the MS for one hundred million years? That would be roughly a B8 star. And what is the largest star that can stay on the MS for a billion years? That is about an A7 star. For our purposes, we shall not consider O and B stars further. The very smallest A stars are conceivably of interest, but the greatest significance must attach to F, G, and later classes of stars because of their vastly greater numbers.

It is among the F stars that the transition from the proton-proton chain to the carbon cycle occurs. By a probable coincidence, it is among the F stars that we also find the transition from very high rotation rates of hundreds of kilometers per second (the norm for O, B, and A stars) to a few kilometers per second (the norm for G, K, and M stars). There has been much speculation regarding the meaning of this break. Some authors have pointed out that, in our Solar System, the total amount of angular momentum is similar to that found in the brighter (upper) MS stars, but that in our system the angular momentum is carried by the planets, not the Sun. It is therefore possible, the argument continues, that the formation of a planetary system occurs with (or is even caused by) the outward transport of angular momentum from the central star. All mechanisms for such transport require considerable time; the evidence from the rotation speeds of MS stars suggests that the break occurs for stars with total MS lifetimes of about two billion years (and therefore with actual ages of between zero and two billion years). The approach that links despinning of the star to planet formation therefore suggests a time scale for planet formation of about a billion years—but this is about ten times as long as it took the terrestrial planets in the Solar System to form. It may have taken that long for the growth of Uranus and Neptune to reach completion, but the connection of that process to the spin rate of the Sun is not obvious.

WHAT ABOUT stars that do not lie on the main sequence? Are any of them stable enough, and long-lived enough, to maintain stable planetary conditions? The best prospect seems to lie in the further evolution of normal MS stars.

When a main sequence star with a mass larger than about 0.6 Suns (roughly G8 on the main sequence) runs low on its hydrogen supply, it can begin fusing helium in the center of its core into carbon. When this new source of energy becomes available, it changes the structure of the star radically, shifting it off the main sequence. The envelope of unburned hydrogen heats up and inflates, sometimes reaching truly enormous size, to become a giant star. From the main sequence star with a diameter of about a million kilometers, it may, depending on its mass and age, reach a diameter in excess of ten million kilometers, giving the star a density of about 0.001 grams per cubic centimeter. (The Sun's density is 1.41, compared to 1.00 for water and 0.001 for air at Earth's surface). A red giant is a red-hot near-vacuum.

During the transition to the giant stage, as the helium burning core lights up for the first time, there is a moment of wild celebration called the helium flash, during which the luminosity of the star briefly increases by a factor of several thousand. The helium flash, even though transient, would have seared the planets with a nearly tenfold increase in temperature. The star then settles down to a brief career as a stable helium-burning star with about one hundred times the luminosity it has while on the main sequence. The location of these helium-burning stars on the luminosity-color diagram is at roughly constant luminosity, with temperatures ranging from white to red, a class called the "horizontal branch" or sometimes the "helium-burning main sequence." Most giants are orange or red in color, leading to the name *red giant* for the class. The effect of a hundredfold increase in luminosity is a rough tripling of the surface temperatures of all planets surrounding the star. Any planet that had previously featured liquid water would, during the giant phase, be broiled at about 1000 K, far beyond the liquid water range, near the beginning of the melting of rocks. If the Galilean satellites survived the helium flash, their ice would melt and deep oceans would cover them. For ten million years (for massive stars) to a hundred million years (solar-mass giants) the star remains fairly constant, burning its helium supply. Here, then, is a possible way to provide Earthlike conditions on worlds orbiting giant stars. A Europa that has long hidden oceanic life beneath a thick ice layer would for a time have its imprisoning ice melted, affording its native life, however advanced it may be, access to air and space. Such hidden life would be well shielded from the violence of the helium flash, which would sterilize the surfaces of normal terrestrial-type planets, but only melt sterile surface ice on the moons of its distant giant planets.

However, the end of the giant phase is less benign. When the helium core of a giant star is depleted, things rapidly get worse. The less massive giant stars heat up and violently expel their remaining hydrogen and helium envelope, exposing their hot, dense cores. They eventually end up as tiny, faint white dwarfs with masses of 1.44 Suns or less. The more massive giants, up to about 8.5 solar masses, generate such high core pressures and temperatures that the fusion process forges ahead, converting carbon into oxygen, neon, magnesium, silicon, sulfur, argon, and calcium. The structural rearrangements of these stars caused by these new sources of energy expand their envelopes even further, a billion kilometers. These enormous, bloated supergiant stars have an average density some one billion times lower than that of the original main sequence star. Planets find themselves actually orbiting inside the hot, tenuous envelope of the star. They evaporate and slowly spiral into the deep interior of the star, where they are ripped apart by tidal forces and dissipated as vapor. Supergiants, not content with broiling their planets, actually expand and swallow them up. The end of the giant era is probably not survivable.

Most damaging, however, are giant stars of about nine solar masses or larger (above main sequence spectral class B3). These stars, after running the course of nuclear fusion up to calcium, continue on to make the iron-group elements, then develop catastrophic internal instabilities and blow themselves up. These supernova explosions hurl up to 90 percent of the mass of the star out at a percent of the speed of light. Supernova explosions briefly shine with the luminosity of a small galaxy. They utterly destroy any distant planets that may remain behind after the supergiant phase. The residue from a supernova explosion, a pulsar or black hole, is a tiny body with incredibly high density. If anything accompanies them, it is the seared core of a former gas giant or brown dwarf that has been stripped of virtually all of its original mass. Even the planets of other nearby stars may be adversely affected.

Clearly, the environments around pulsars, black holes, and white dwarfs, the postgiant residues of stellar evolution, have been sterilized by the prior evolutionary history of their central star. Although certain niches near red giants are stable and benign for periods of many millions of years, it is far from clear that life could evolve to a high level during such a brief interval of time. It is also unlikely that any niche other than Europa-like bodies could have had conditions conducive to life both before and after the transition from main sequence to giant. But, for a Europa to maintain a deep ice-covered ocean until the "giant spring" thaws out its ice layer, that body must have been in a close orbit around a gas-giant planet in the company of other massive moons. Tidal heating in such a system is the only way to keep the ocean from freezing. Another difficulty is imposed by the searing luminosity of the helium flash. Again, only a Europa, with its thick surface layer of ice, might sustain that heating episode without fatal damage to its ocean-floor ecology. But even a Europa with deep water oceans, orbiting with its gas-giant planet around a small red giant star, faces a grim future: at the end of the relatively brief giant phase, the star will destroy its system utterly, either by violent mass ejection or by running out of helium fuel and falling into the relative oblivion of white dwarf status.

WE HAVE now explored the benefits and hazards of a wide range of spiral-arm stellar environments for various classes of planets. But, as the puzzled reader has surely noticed, we have not reached the end of the chapter. It is unfortunately true that having an understanding of every kind of planet and every kind of star is not sufficient to an understanding of the Universe. The reason is simple: most stars are not single. As many as two-thirds of all stars are members of double, triple, or higher-multiple stellar systems, in which stars orbit each other. How do the mousy little planets fare during the dance of the elephants? It takes only one misstep in the choreography of a complex system to sterilize or destroy a life-bearing planet.

Multiple stars can exist with similar or dissimilar masses, and can be located at almost any distance from each other. Some doubles are essentially in contact with each other, with gases flowing from one star to the other. Others are so far apart that it takes hundreds to thousands of years for them to complete an orbit. Some have highly eccentric orbits, in which the separation of the stars is usually so great that they have little effect upon each other, but in which the rare, swift perihelion passages may be extremely disruptive to the planets of either star.

In principle, very close binaries may be compatible with the existence of planets far from the center of the system. There are, however, certain important limits on close binaries. First let us consider close binary stars in which the two members are of the same mass, luminosity, and spectral type. Two close M7 dwarfs, each with a luminosity of 0.0001 Suns, would support Earthlike conditions on a planet at a distance of 0.014 AU. The two binaries would have to be about 2.5 Jupiter radii apart for an orbital period of about one hour to avoid tidally stripping each other. The two stars would take turns eclipsing each other every half-hour. The planet with Earthlike temperatures, at 33 Jupiter radii (0.014 AU), would have an orbital period of about four hours. This is too close to the star pair for comfort. But close binary systems with earlier M stars would have more luminosity, so that the planet could be set at a distance that was not only thermally appropriate, but also dynamically safe.

Systems containing two unequal-mass stars behave in very much the same way if the masses of both stars are less than 0.6 Suns. But there is an interesting potential for mischief when the mass of either star passes 0.6 Suns. The more massive star will, according to the principles we have discussed, evolve more rapidly than its smaller sister. Above a mass of 0.6 Suns, that star will eventually evolve off the main sequence into the giant phase, expanding enormously as it does so. In close binary systems, the gaseous envelope of the giant will fall onto and be captured by the smaller star, pumping up its mass. Since the outer envelope of a giant is unburned hydrogen that was too close to the surface of the star to experience fusion, this mass transfer process essentially promotes the smaller star up the main sequence to a higher mass (and shorter lifetime) by adding fuel to its envelope. Meanwhile, the removal of mass from the larger, more highly evolved star lowers its internal pressure and temperature and slows its evolution. Transfer of large masses of gas often triggers instability in the receiving star, causing brief, violent outbursts called nova explosions. Such explosions would be disastrous for planets in the system.

The threshold for this behavior, 0.6 solar masses, corresponds to an orange-red K6 star with a main sequence lifetime of fifty billion years. Since the Universe is only about twelve billion years old, no K6 star has yet reached the end of its MS lifetime. But a star with an MS lifetime of less than twelve billion years could have already reached the point of departure from the MS. That corresponds to a first-generation G7 star with about 90 percent of the mass of the Sun. Stars of this mass or larger in close binary systems are capable of initiating mass transfer and causing nova explosions. Thus for close binary stars, we could accommodate safe planets if neither star is bigger than G7, but at least one is larger than about M5.

Clearly it is disadvantageous to have a massive star in a close binary system. It shortens the expected lifetime of both stars and limits the useful shelf life of all planets in the system.

However, distant binaries can easily be constructed to allow stable orbits around one or both stars. For example, an M-class star of 0.1 solar masses (one hundred Jupiters) would have about the same gravitational effect on Earth as Jupiter does if it were ten times as far away. That would be 50 AU from the Sun, safely beyond the orbit of Pluto. The M star's orbital period of 350 years could be in a near-resonance with Neptune, but the inner planets would generally experience no direct effects. The luminosity of the star would be completely negligible even on Neptune. Circular orbits at greater distances, or eccentric orbits that approach the Sun no closer than 50 AU, would be equally satisfactory. The presence of such a star during the era of planetary formation might have kicked more cometary material (that which presently resides in the Kuiper belt) into high-eccentricity orbits where they could both add precious volatiles to the terrestrial planets and threaten them with mass destruction.

Planetary systems orbiting less massive central stars would be somewhat more sensitive to disturbances. For example, a central star of mass 0.25 Suns would require exiling the secondary star to 100 AU or more. Conversely, more massive central stars would be even more forgiving. Distant binaries seem to present no great problems. Indeed, in many cases both stars would be capable of maintaining planets in stable orbits.

Then, considering all these factors, what stars should we seek when considering starting life on a planetary surface? The favored choice is a solitary main sequence star belong to spectral classes from F0 to about M6, but very close or very distant doubles involving only these classes of stars are also often acceptable. Stars that do not belong to the main sequence are so transient that they are poor choices for the origin of life. However, giant stars with distant Europas and preexistent sub-ice life are an interesting exception.

14

STARRY NIGHTS: LIFE IN A GLOBULAR CLUSTER

NOW THAT WE HAVE BECOME familiar with life in a spiral galaxy, the time has come to seek a change of scene. In the Universe, about half of all the stars exist in elliptical galaxies that differ in every way from spirals. Spirals are flattened, lenslike in overall shape, with pronounced spiral lanes of stars and dust. Large spirals frequently have more than one hundred billion stars. The spiral lanes are marked by strings of very dense, collapsing interstellar clouds, star-forming regions, and their accompanying highly luminous upper main sequence stars and supernovas. By contrast, the elliptical galaxies are highly three-dimensional swarms of stars, elliptical or circular in outline, with little or no gas and dust. They show no signs of present or recent star-forming activity, no short-lifetime upper main sequence O, B, A, or F stars, no T-Tauri stars, few giants, and no supernovas. The very stars that are most hostile to life are lacking.

Elliptical galaxies are all so remote from us that detailed observations of them are very difficult. But, fortunately, the Milky Way is accompanied by a great spherical swarm of globular clusters of stars that look like tiny elliptical galaxies, each containing thousands to hundreds of millions of stars. Further, they are free of gas and dust, and the stars in them are closely similar to those in ellipticals.

We have already explored the familiar, nearby population of galactic spiral-arm stars in some detail. These stars, the first discovered (by naked eye in prehistoric times) and the first studied in detail, were thought to be typical of stars everywhere until systematic studies of the spectra and color distributions of stars early in this century showed otherwise. When it became clear that the members of globular clusters and elliptical galaxies differed profoundly from the familiar nearby population of stars, a distinction was made between Population I (nearby, spiral-arm stars) and Population II (globular cluster and elliptical-galaxy stars).

Population II stars contain hundreds of times less heavy elements than Population I stars. Since the heavy elements are made by nuclear-fusion reactions in the interiors of massive stars, and since they are ejected violently into the interstellar medium by the catastrophic supernova explosions of massive stars, the general tendency of the universe is for stars formed early in history to have a composition very similar to that of the debris from the Big Bang explosion (virtually pure hydrogen and helium). Stars formed later by the collapse of interstellar gas and dust clouds are enriched by the heavy elements created and expelled by earlier generations of stars. The most recently formed stars contain the highest concentrations of heavy elements such as carbon, oxygen, silicon, and so on. These are the ice-forming and rock-forming elements. Therefore a Population II assemblage of stars in a globular cluster orbiting the center of the Milky Way galaxy, containing four to forty times lower concentrations of the heavy elements, must consist of stars that were formed at a very early stage of the chemical evolution of the Universe.

When we look at the distribution of Population II stars on a luminosity-color diagram and compare them to Population I stars, we at once notice several other striking differences. First, the entire upper main sequence is missing from Population II. In many cases, no stars brighter than spectral class G4 are present on the main sequence. From simple calculations based on the supply of hydrogen fuel in these stars and their luminosity (rate of fuel consumption), we can conclude that such a cluster must be old enough for all the G3 stars (and larger) to have evolved off the main sequence. We know that a G4 star has about 90 percent of the mass of the Sun, a luminosity of about 70 percent of that of the Sun, and a main sequence lifetime of about 13 billion years. Applying this method to a large number of other globular clusters, almost all give ages between 10 and 15 billion years. If the rate of production of heavy elements has been constant since the first appearance of stars, then the existence of stars with forty times lower heavy-element abundances than today's new stars implies that they were formed one-fortieth of the way through the history of the Universe. This idea is confirmed by the observation that the most recently formed stars contain heavy-element abundances that are more than 40 percent higher than we have in the Sun. Given an age of 4.6 billion years for the element mix in the Solar System, as determined by several kinds of radioactive decay dating, these compositions again indicate an age of about 12 billion years for the Universe.

There are relatively few giant stars in the globular clusters, and these are of modest mass, consisting of G-type stars that have recently turned off the main sequence. It is in a way sobering to look at a stellar population in which there are no longer any stars like our Sun, because they have all expired from old age! There are also no supernova explosions in these clusters. In a region in which there has been no star formation for at least 10 billion years, all those stars massive enough to be capable of supernova explosions (at least nine solar masses; main sequence lifetimes of 7 million years or less) must have blown up long ago. In fact, they would have been cleaned out of the globular-cluster population after only 0.04 percent of the present age of the cluster.

There is another, less meaningful difference between the observable properties of globular-cluster stars and nearby Population I stars: even the nearest globular cluster is so far from us that stars with very low luminosities cannot be seen. This means that the two important classes of low-luminosity stars, the lower (M-class) main sequence stars and the white dwarfs, are not seen in globular clusters. But theory tells us that both should be abundant. If we were close enough, we should see lots of both types. We have every reason to believe that it is only observational limitations that keep us from seeing them.

What would it be like to reside in a globular cluster containing a million stars? Certainly, the sky would be very busy at night. There would be, it is true, no spectacular white and blue O, B, and A stars of enormous luminosity. But the globular cluster, even though made up of modest yellow, orange, and red stars, is built on quite an intimate scale. The nearest stars are much closer to each other than we are accustomed to in our location in a spiral arm. The cluster also contains no dense, dark dust clouds to obscure the sky. In our own galaxy, recognition of the spiral structure was delayed until the last century by the heavy obscuration of its structure caused by nearby dark clouds ("coal sacks"), but in a globular cluster, the entire symmetrical architecture would be evident to any observer.

A million-star cluster would, according to the basic theory of the aging of single stars, be dominated by M and K stars: about 730,000 would be red M stars and 150,000 would be orange K stars. Roughly 100,000 would be the white dwarf residues left by departed upper main sequence stars, and a paltry 20,000 would be late G stars of classes G5 through G9. From outside, we would describe the cluster as faint and subdued.

But the naked-eye observer within the globular cluster would form a rather different image of his surroundings. He would see a grand total of 120,000 stars in the sky (assuming that he could see down to an apparent magnitude of 6, which is about the average limit for humans, who see a total of about 3000 stars in Earth's skies). Of these, about 16,000 are brighter than magnitude 3 (compared to 152 for Earth), 3700 are brighter than magnitude 1 (versus 13 for Earth), and 300 are brighter than magnitude −1 (only Sirius appears this bright as seen from Earth). The observer would notice that all the stars brighter than −2 magnitude were yellow, and all those brighter than magnitude 1 are orange or yellow. Of all the visible stars, some 20,000 would be yellow G stars (every one in the cluster would be luminous enough and close enough to be visible), 94,000 would be orange K stars (another 56,000 would be too faint and distant to be seen), and only 6700 would be red M stars. Astonishingly, fewer than 1 percent of the M stars even in this extremely compact cluster would be close enough to be seen by a naked-eye observer. In fact, the odds are that not one single M8 or M9 star of the 230,000 present would be close enough to be visible. Of the white dwarfs, only about 200 would be close enough to be visible as tiny, faint white stars.

But there is another class of stellar bodies in globular clusters that are not predicted by the theory of single-star evolution. The Hubble Space Telescope has found numerous faint blue stars in the hearts of large globular clusters. There is only one way known to make such stars: by mass exchange between members of close binaries. Picture two stars of, say, classes G5 and M2, in close (but possibly very eccentric) orbit. The G star, with a lifetime of eleven billion years, now begins to evolve off the main sequence and enter the giant stage. But, as it inflates, the hydrogen-rich gas from its outer envelope is captured by the red dwarf. The removal of atmosphere dampens the fires of the G star and feeds the flames of the M star. The G star may be prevented from evolving into the giant region by mass loss, losing its outer atmosphere completely. The former M star, now fanned into a burst of hyperactivity, is no longer a simple, innocent, main sequence star, but a new beast of strange structure and history. Such stars are prone to episodic, moderately violent explosions called novas. We mentioned earlier that supernova explosions do not occur in globular clusters. But doesn't mass exchange between stars permit them to build up to the point of instability? The answer is that the largest stars present have masses of about 0.8 Suns, whereas it takes about 9 Suns to make a supernova. Big-time theft is impossible where everybody is poor.

Why should close binaries be numerous in globular clusters? The answer is twofold: they are important in globular clusters because they are important everywhere; and they are much more readily made by close stellar encounters and gravitational capture in globular clusters because the number of stars per cubic parsec is so enormous (several thousand, versus less than one in spiral-arm space near the Sun) that close encounters are common. Marriage is convenient. Not only is it easy to capture a star in orbit around another, but it is also easy to strip a loosely bound binary apart. Marriages of convenience don't last.

In the heart of the globular cluster, each star passes within 200 AU of another star about once in every ten million years. On the average, such an encounter happens somewhere in the cluster about once every ten years. Over the fifteen billion years of history of the cluster, each star will suffer over one thousand close encounters. At relative speeds of about twenty kilometers per second, the typical close flyby will last several years. Each such pass can induce an eccentricity change of about 0.1 in the orbit of a distant planet located 70 AU from one of the stars. The very closest pass of the thousand events would be at a distance of few AU, capable of disturbing the orbits of even close planets. In the long run, as these disturbances accumulate, loss of distant planets seems likely.

Given that the density of stars in space is about ten thousand times higher in a globular cluster than in the spiral arms, is there a possibility of collisions of stars occurring? Rough calculations suggest that the probability of a single collision occurring in a globular cluster containing a million stars over twelve billion years is less than 0.1. But a large spiral galaxy with two hundred globular clusters may expect about one such collision every billion years.

The typical globular cluster pursues an elliptical orbit around the center of the Milky Way, completing each circuit in about two hundred million years. Twice on each orbit the cluster dives through the plane of the spiral. Over twelve billion years, a cluster makes some 120 crossings of the galactic plane. Rather frequently, perhaps once in every few dozen plane crossings, it must pass through regions rich in gas and dust at its relative speed of about 240 kilometers per second. At that speed a dust grain would have a kinetic energy equal to the explosive power of over one hundred thousand times its weight of TNT. The entire cluster is swept clear of any gas, and every planet in the cluster is bombarded with spiral-arm hydrogen gas. A hydrogen atom traveling at 240 kilometers per second has an effective temperature of 4 million degrees K, high enough to drive some low-energy nuclear reactions. The impact of these atoms would generate a flood of deadly X rays.

In the course of stellar evolution, many members of the cluster must have passed through massive gas-loss episodes, or even suffered supernova explosions. Since no stars younger than about ten billion years are found in most of these clusters, all massive stars that ever existed in the cluster must have been formed at least ten billion years ago. Massive supernova precursors have main sequence lifetimes of only thirty million years or less, so any supernova explosions that occurred in the cluster must have been very early in history. But a cluster is so compact that a supernova explosion would almost certainly strip it of its gas and dust. In other words, even neglecting the stripping of gases from a globular cluster by passage through the spiral arms, the raw materials needed for star formation must be quickly lost from the cluster: star formation must cease at an early date.

Passage through the galactic plane also entails a significant risk of encounter with massive spiral-arm stars. The core of the cluster will sweep out a cylindrical volume through the galactic plane that will contain several thousand spiral-arm stars, of which several will be B stars with masses of about twenty Suns. The closest encounters will be at distances of a few thousand AU, but gravitational disturbances of the orbits of distant cluster members may be sufficient to detach them from the cluster and launch them into independent high-inclination orbits around the galactic core. Although such a fly-through of the galactic plane sounds frightening, the probability of a collision between a globular-cluster star and a spiral-arm star is negligible.

The internal dynamics of the cluster, with many fairly close encounters, favor the slow approach to a state in which the component stars have equal kinetic energies. Such a state of "equipartition of energy" imposes high velocities on the smallest stars, making it far more likely that an M8 star of 0.09 solar masses will escape from the cluster than a late G star with 0.8 solar masses will do so. In fact, the smaller star will acquire three times the velocity of the larger. Thus M stars may not only be hard to see in globular clusters on account of their small mass: they may actually be selectively ejected from the cluster for the same ultimate reason.

All this does not sound like a congenial environment for planets and life. Actually, things are quite a bit worse than the dismal picture painted so far. The greatest problem is that the raw materials of the cluster are so ancient that the heavy elements are very rare, ten to a hundred times less abundant than in our spiral-arm neighborhood. The heavy elements, meaning oxygen, carbon, nitrogen, magnesium, silicon, sulfur, calcium, aluminum, iron, and so on — all the elements essential for making terrestrial planets, rocks, ices, and life — are in very short supply. Solid bodies of ices and rocks must be either far smaller than their counterparts in our Solar System or there must be far fewer of them. The important radioactive heat sources potassium-40 and uranium-235, even if present initially with about the same concentration they had during Earth's formation, must have decayed away almost completely: after twelve billion years, only 0.4 percent of the original amounts of these materials would remain undecayed. Their concentrations in rocky bodies would today be fully thirty times lower than in the terrestrial planets of our own Solar System. Geological activity cannot continue in small bodies with such feeble heat sources. Any rocky planets in a globular cluster must by now be geologically dead.

Even normal gas-giant planets may not form, since present theory suggests a crucial role for gas capture by large ice-plus-rock bodies, sometimes affectionately referred to in the trade as "mudballs." Gas-rich bodies formed by rotational instability, including brown dwarfs and possibly some extra-large gas giants, should not be affected by this raw-materials problem.

Even once we have brown dwarfs and giant gas giants, there is still a problem making satellites for them. A body of Europa-like composition but thirty times' smaller mass would probably not even melt and differentiate into ice and rock layers. A family of mini-Galilean satellites thirty times less massive than Jupiter's big four would not transfer enough tidal energy to keep their ice crusts melted even if they did somehow differentiate. No Earths and no Europas suggest no decent niches for the synthesis and accumulation of organic matter and the origin of life.

The last refuge of pro-life sentiment in globular clusters must be the brown dwarfs themselves. Despite their slow shrinkage and cooling, there are liquid-water zones in the atmospheres of the smallest brown dwarfs. Those larger than about fifteen Jupiter masses are simply too hot; planets in the range of masses from Saturn (0.3 Jupiters) to about three Jupiters may have and preserve liquid-water zones. But the classical objection to the origin of life in such bodies still applies: the instability of the atmosphere. Convective overturn, the heat engine that delivers heat to the top of the atmosphere and dumps it into space to enable further shrinkage to occur, rapidly transports material from benign temperature regions to either killingly cold high altitudes or killingly hot low altitudes. While life is not categorically impossible in such an environment, it certainly looks like a long shot.

Brown dwarfs, however, do have another advantage: they are astonishingly autonomous bodies, not reliant on the presence of a central star. In globular clusters, autonomy is to be treasured: any planet may at any time find itself cast loose from its parent star and launched into looping orbits within the cluster, or even fired out of the cluster altogether by strong gravitational interactions with other stars. Most planets would perish at the thought of finding "Destination: Void" stamped on their tickets. Not so for brown dwarfs.

15

INTERSTELLAR ORPHANS

WHEN, IN THE COURSE OF cosmic events, it becomes necessary for one planet to dissolve the gravitational bonds that have connected it with its mother, and to assume among the bodies of the Galaxy the separate and equal station to which the laws of Nature and of Nature's God entitle it, a decent respect to the curiosity of mankind requires that we state the causes that impel it to that separation. It would also be helpful to explain what happens as a result of separation.

In the previous few chapters we have encountered only two types of environments that are compatible with the long-term presence of liquid water in the absence of a star. These are the atmosphere of a gas-giant planet with a mass between about one-third and three times that of Jupiter, in which the internal heat source maintains a liquid-water cloud layer, and the interiors of large icy satellites kept warm by tidal forces.

The largest of these gas-giant planets approaches the lower end of the brown dwarf size range. Because of the slow shrinkage and cooling of gas giants, their "surface" temperatures slowly decline with time.

The shrinkage and heat loss of a gas-giant planet, and the consequent cooling of its visible surface, means that such a planet always starts out too warm for water clouds. For gas-giant planets, the surface temperature is that observed from space: for Saturn or Jupiter, it corresponds to the temperature of the topmost cloud layers. Cooling eventually allows water vapor to condense at the coolest point in the planet's troposphere. That point is the boundary between the convective, turbulent troposphere and the thin, cool stratosphere. The total pressure (and the water-vapor pressure) are so low at that altitude that, when temperatures first get low enough for water to condense, it is ice crystals that form, not water droplets. For a body the size of Jupiter, early surface temperatures are as high as 1000 K a million years after the beginning of planetary collapse. Cooling to a "surface" temperature of about 450 K, just cold enough to condense a thin high-altitude haze of water frost, takes about 4 million years. From that time until about a billion years have elapsed, when the surface temperature of the Jupiter-like planet has dropped to about 140 K, there is an ice-cloud haze that steadily increases in density as its base extends deeper into the atmosphere, to ever higher pressures. At a planetary surface temperature of about 140 K, the base of the water condensation region finally drops to levels where the temperature is higher than 273 K, and liquid water droplets become stable at the base of the ice cloud. Freeze-thaw cycling of cloud particles becomes important for the first time, bringing with it the generation of electrical charges and the possibility of large-scale lightning activity. Almost simultaneously, high-altitude clouds of crystalline solid ammonia begin to form near the tropopause. During the ensuing 3.6 billion years, right up to the present, liquid water droplets remain stable in Jupiter-size planets. They will continue to be stable until the Sun departs from the main sequence and enters the giant phase at about ten billion years age.

Saturn-size bodies start cooler and cool more quickly than Jupiters. Their liquid water clouds would form earlier and persist longer.

Super-Jovian planets are all less hospitable to water. On a body with three times the mass of Jupiter, water-ice condensation does not begin until 50 million years after formation, and water-droplet clouds do not become stable until about 4.5 billion years: a planet of three Jupiter masses in our Solar System at Jupiter's distance from the Sun would only now be evolving to the point where liquid water clouds would be stable. Even more massive planets would take so long to reach water stability that all main sequence stars except K and M stars would die of old age before liquid water appeared. For example, planets with ten Jupiter masses would take about 20 billion years to cool to the point at which liquid-water clouds become possible. The Universe is not old enough for such bodies to have appeared. Bodies of that size are on the threshold of deuterium fusion, and are better described as brown dwarfs than gas-giant planets.

On Population II gas giants, however, the abundance of water (and just about everything else of interest) is at least several times lower than in solar material. This causes a problem. To get this very dry gas cold enough to condense as clouds, the temperature must be brought down to below the freezing point of water. Water vapor in a rising, cooling column of gas would first condense as ice particles. At higher altitudes in the clouds the temperatures are even lower. By the point at which ammonia begins to react with the ice clouds to make highly concentrated ammonia solutions, the temperature is so low that almost no water vapor is left. Tiny amounts of very cold, very concentrated solutions of ammonia in water are produced, making tenuous clouds ten thousand times less dense than the low-lying snow clouds. That "liquid" solution is a very strange material, so cold that it is extremely viscous, more like glass than liquid water. Up-end a cup of this solution and it could take days to pour out. It has the appearance and structure of glass, an apparent solid without crystal structure; in fact, technically it is a kind of glass. These glassy droplets cannot simply be dropped into lower, warmer levels of the atmosphere to melt. They are sufficiently volatile, because of their high ammonia content, so that they would quickly evaporate into ammonia gas and solid ice, and then the falling ice itself would completely evaporate well before reaching the melting point. In a Population II gas giant or brown dwarf there is actually no liquid phase in which organic compounds could dissolve, accumulate, and react. The road to life is unpaved—and covered with ice.

There is only one way to force a water-poor (Population II) gas giant or brown dwarf to bring forth liquid water clouds: move the condensation point down to high pressures. This is equivalent to cooling the entire atmosphere, or to lowering the surface temperature. The way to achieve this feat is to find a planet with a surface temperature of 100 K or less. In other words, the best conditions in Population II gas-giant planets are found in the smallest planets, far below their cloud tops. The Uranuses and Neptunes of such systems, if they exist, would work nicely. The problem is that the formation of Uranuses and Neptunes, at least according to current theory, depends on the presence of abundant ices in the preplanetary nebula from which they formed. Population II Uranian planets may very well not exist. True, one might conceive of planets with the mass of Uranus and very small core masses of ices and dust—essentially, small, hydrogen- and helium-rich Jovian planets. But the time required to build such planets by core accretion and subsequent gas capture may be longer than the entire lifetime of a prestellar nebula.

The raw materials could easily be dissipated before the job of planet assembly was well begun.

BROWN DWARFS can play a useful role in maintaining biologically interesting temperatures on bodies orbiting them. They can do this in two fundamentally different ways, either by radiant heating of satellites with liquid-water surfaces (Earths and Earthlets), or by tidal heating of bodies with ice-covered deep oceans (Europas). Tidal heating may be effective around planets with masses as low as 0.1 to 0.3 Jupiters: for the present discussion, we will lump Jovian planets together with brown dwarfs in discussing tidal heating.

Is it reasonable to postulate a body with roughly Earthlike composition in orbit around a brown dwarf? The closest thing to a brown dwarf in our Solar System is Jupiter. The question should then be rephrased, Is there any body of Earthlike composition in orbit around Jupiter? The answer is no — sort of. Actually, Io is at present a very dry, highly evolved body. At the time of the *Voyager 1* Jupiter flyby, when Io's vast sulfur flows and violent volcanic eruptions were confirmed by stunning spacecraft imagery, I proposed an explanation for the unique geochemistry of Io (actually, with tongue firmly in cheek, I referred to it as Iochemistry). There was apparently only one route to get to a planet with vast quantities of sulfur and sulfur dioxide on its surface and no detectable water. Io had to start with sulfide minerals and a lot of a mild oxidizing agent. That oxidizing agent was almost certainly water. Melting and differentiation of Io must have released water to its crust and surface via volcanism. At the temperature of a volcanic vent, a significant fraction of the water vapor emerges as hydrogen gas, which promptly escapes from little Io and gets lost in space. The oxygen ties up with sulfur from the sulfide minerals, making sulfur dioxide. After many cycles, the hydrogen is all gone and the oxygen from the water is all tied up in iron oxides and sulfur dioxide. The predicted original composition for Io was very similar to that of Mars. The big difference was that rapid, energetic, persistent recycling of crustal materials on Io drove the reactions and escape process to the point of completion. Water was exhausted. Mars, by comparison, has no continental drift and precious little recycling. Why?

The answer to this question was supplied one week before the question was asked by an astonishingly prescient paper written by Stanton J. Peale of the University of California at Santa Barbara and Pat Cassen and Ray Reynolds of NASA Ames Research Center. They pointed out that the four Galilean satellites actually all had small orbital eccentricities. Their mutual gravitational perturbations tended to pump up each other's eccentricities (much in the manner of a human mob), producing results that none of these bodies would have been capable of alone: since all the Galilean satellites are deep within Jupiter's gravity well and rotationally locked on to Jupiter in 1:1 spin-orbit resonances, each always keeps one side (the end of its long axis) perpetually pointed at Jupiter. But, because their mutual perturbations give them all some orbital eccentricity, it is actually impossible for them to keep their tidal bulges perfectly lined up toward Jupiter. After all, their rotation rates are constant, but their rates of travel around their orbits vary periodically: the peak of the tidal bulge swings through a few degrees back and forth about the direction of Jupiter with every orbit. The tidal forces of Jupiter actually distort the crust, bending it back and forth, up and down, as Io slowly waggles back and forth. Io, being deepest within Jupiter's gravity field, feels by far the strongest tidal forces (they drop off with the cube of distance from the center of Jupiter). Like a metal bar being repeatedly bent back and forth, the energy that goes into flexure of the crust is partly transformed into frictional heat. Io gets hot inside. Europa, a little farther out, gets a milder version of the same treatment. So, in the final analysis, Io is a sulfur planet because it lost all its water, because it is so good at recycling crust, because it has an internal heat source, because it is locked into strong tidal interactions with Europa and Ganymede. It turned out different from Mars because of peer pressure.

The secret to getting a Mars or Earthlet to survive in orbit around a brown dwarf is that peer pressure must be absent or negligible. A lone Ganymede-size satellite (or an Earthlet, Mars, or Titan) would not lose its water. For any brown dwarf with an adequate surface temperature it should be possible to have a planet of this size in 1:1 resonance, its tidal bulge firmly and permanently locked on to the brown dwarf, with zero orbital inclination and eccentricity. Over billions of years the satellite would slowly recede from its primary. The brown dwarf also must constantly cool off and shrink to ever smaller radius and ever higher density.

The combination of slow retreat of satellites and slow cooling of the brown dwarf creates a problem. Once conditions suitable for life are achieved on the surface of the satellite, how long might we expect them to last?

For a small brown dwarf (about thirteen Jupiter masses, at the onset of transient deuterium burning), the surface temperature after a billion years of cooling is about 530 K, still too warm for water- or ice-bearing clouds. After about five billion years the temperature is down to 280 K. The brown dwarf has during this time shrunk from about 76,000 km to 72,000 km radius. Its surface gravity, about thirty Earth gravities, also slowly increases during this time. Suppose we try to place an Earthlet at a distance from the 530 K body such that it will have a temperature of 300 K. That distance works out to 135,000 km. Unfortunately, that is only 1.77 times the radius of the brown dwarf, well within the Roche limit for tidal disruption of the planet. Since this brown dwarf has a density of 15.4 grams per cubic centimeter, it can disrupt a planetary body with a density of 5.5 (Earth) out to a distance of over 260,000 km. The devastating tidal forces definitely rule out any Earthlike planet around a minimum-sized brown dwarf, even one as young as a billion years. Older, cooler, minimum-mass brown dwarfs are obviously even worse.

More massive brown dwarfs are denser and hotter: the higher density makes their gravitational attraction even more destructive, but the higher temperatures allow more distant bodies to have warm temperatures. Which is the dominant effect?

A maximum-sized brown dwarf (0.075 Solar masses, or about fifty-four Jupiters) with an age of a billion years has a radius of 66,000 km and a temperature of 2300 K. Earthlike temperatures are found at a distance of a healthy 2.2 million kilometers. The density of this maximal dwarf (a new record: the Universe's biggest brown dwarf!) is an astonishing 140 grams per cubic centimeter, and the Roche limit is out at 480,000 km. The planet is therefore completely safe from tidal destruction, but it is still so close that it must be in a resonance lock with the brown dwarf.

But conditions at any point near the brown dwarf must change with time, reflecting the cooling of the dwarf. The very same brown dwarf, after three billion years of aging, has shrunk and cooled to about 1600 K. The ideal distance for a 300 K planet would then be about 1.87 million kilometers, well inside what was the ideal distance two billion years earlier. From the point of view of the once-warm planet, somebody has turned down the thermostat. Its oceans would easily freeze over.

The best way for a planet orbiting a brown dwarf to accommodate itself to the cooling of the dwarf is for temperatures on the planet to have been close to the boiling point of water early in history, declining toward the lower end of the liquid range as time passes. Such a history can provide billions of years of liquid water on a planet at the right distance from the brown dwarf.

There is another possibility: the planet may indeed cool to the point that it freezes over after several billion years. Then we would cease to call it an Earthlet and start calling it a Europa. The shift from surface-correlated life to ocean-floor vent life would be devastating to Earth, because most life on Earth is dependent not only on solar surface heating but on photosynthesis. But photosynthesis as we know it is not an option on a planet orbiting a brown dwarf. Not only is the illumination feeble; it is devoid of photons energetic enough to drive photosynthesis. In this case, the bottom of the food chain is not photosynthetic green planets, but inorganic or organic chemical potential energy fed in by the nutrient-rich broth emanating from the hydrothermal vents. Note that we are not assuming that a planetary ecosphere once powered by a sun could adapt readily to the loss of sunlight and learn to feed from ocean-floor vents: we are considering worlds on which sun power was never an important factor, and life has always fed from hydrothermal vents.

This discussion of using the radiant heat from brown dwarfs to keep liquid water stable on the surfaces of nearby planets has again led us back to the idea of Europa-like bodies with deep, often ice-crusted, oceans. The maintenance of liquid oceans depends only on the continuing presence of tidal-energy dissipation. In turn, tidal heating requires that there be, as in Jupiter's satellite system, at least two massive bodies in orbit around the central gas giant or brown dwarf. Earthlet-like bodies may, as we have already discussed, cool and evolve into Europas. Alternatively, surface melting may never have occurred, or was so early and of such limited duration that it had no long-term significance. In effect, the distinction between Earthlets and Europas in a brown dwarf environment has become fuzzy to the point of confusion.

The idea of a life cycle on a planet that cannot depend on photosynthesis sounds bizarre to human ears, but there are ample precedents on Earth. The ocean-floor vent communities are an obvious example of declaring independence from the Sun. But there are whole classes of microorganisms that can derive their energy from inorganic reactions. Such organisms, called autotrophs, apparently arose on Earth early in history, before the rise of oxygen in the atmosphere. These cells neither made nor required abundant oxygen. On Earth, the first autotrophic cells must have been preceded by simple heterotrophic cells, which depended upon fermentation (partial oxidation) of prebiological organic matter for their source of energy. Preexisting organic matter seems essential in any case; therefore, the sequence of early life forms on a Europa may have been similar to that on Earth.

Oxidation and reduction processes may involve several elements besides carbon. Life on Earth uses phosphorus compounds. Iron is another good oxygen carrier, and many terrestrial life forms make extensive biological use of the different oxidation states of sulfur. But whatever elements or compounds are enlisted to transport oxygen and regulate oxidation, the fundamental element of life as we know it is carbon. In Chapter 17 we shall search for alternative chemistries that might broaden our definition of habitable worlds to include others where organic chemistry and liquid water are absent.

Natural inorganic processes in hydrothermal vents and sub-ice oceans can generate hydrogen or oxygen. Dissolved radioactive materials, especially potassium-40, irradiate water with high-energy electrons and gamma rays, tearing apart water molecules to make hydrogen and oxygen gases and liquid hydrogen peroxide. These gases may accumulate in solution in water at high pressures and low temperatures. They may also be released as bubbles when water is exposed by impacts or extruded onto the surface. Hydrogen may escape readily, but oxygen is likely to be retained. An internal source of organic matter and an external accumulation of oxygen make surface life possible even if the surface is frozen, as in Antarctica. Photosynthesis is not essential as a source of oxygen if radiolysis and outgassing are present. But the rate of supply of oxygen from radiolysis cannot be larger than 1 percent of the rate of oxygen production by photosynthesis or photolysis.

In summary, Population I brown dwarfs may successfully host life-supporting Europas, ice-free Europoids, quasi-Earths, or Earthlets. Population II brown dwarfs are so deficient in heavy elements that building satellites large enough to be thermally and chemically interesting would be very difficult.

How would you know from external observation that life existed inside an alien Europa? First, watch out for freeze-dried fish in orbit. Second, sample the freshest surface ice veins and analyze them for organic matter and microfossils, such as bacteria. Third, land on the surface and drill down into the ocean.

WHATEVER FORM of cosmic vehicle we use to transport life through sunless space, whether gas-giant planets or brown dwarfs or Earthlets and Europas orbiting them, there are certain factors these environments have in common. First, there is essentially no visible light. The greenhouse effect and photosynthesis can be forgotten. Second, there is no ultraviolet light. There is no UV-driven photolysis of water, methane, and ammonia, and no escape of hydrogen made by these reactions. Prebiological chemistry must therefore be powered by interior thermal energy, lightning, or radioactivity. Third, the absence of UV photons energetic enough to ionize molecules means that the dissociative recombination mechanism cannot be at work stripping carbon, nitrogen, and oxygen atoms off small planets. Fourth, the absence of UV irradiation means low exospheric temperatures on the planets, inhibiting thermal loss of even the lightest gases. Fifth, these compact, low-mass systems will have very low impact rates. Any unaccreted material left over after the formation of these compact satellite systems will be swept up or ejected in a tiny fraction of a billion years. Explosive blowoff of atmospheric gases by large, violent impacts will not be important.

The consequences of the absence of UV and impacts are profound: it means that a Mars-mass Earthlet (or Mars itself) could do an excellent job of retaining its original endowment of volatiles for many billions of years. Even escape velocities of only four kilometers per second, intermediate between those of the Moon and Mars, may suffice to keep the precious volatiles from escaping. The biggest environmental problem in such a system is maintaining temperatures within the liquid-water range. For intelligent life trapped under ice in such an orphaned system, there would also be the hazards associated with experiencing the ultimate feeling of alienation.

Terrible things can happen to a planet that's not being careful. Some of these fates are reserved for planets that go out into the night alone in bad galactic neighborhoods, but others, equally traumatic, can visit the unsuspecting planet at home. There is no such thing as full security.

16

A PLANET'S LOT IS NOT A HAPPY ONE

A HOST OF PROBLEMS MAY ARISE to threaten a populated world. Some of these are due to evolutionary processes on the planet, such as those involving loss or depletion of atmospheric gases. Some are caused by dynamical events such as changes in the spin or orbital properties of the planet. Some involve violent collisions with other planetary bodies. Some are caused by the slow evolution of the parent star. Some are "cosmic accidents" such as close encounters with another star or a nearby supernova explosion. Fortunately, most of these are either very rare or predictable far in advance. But, as the proud owners of a middle-aged planet, we should be aware of what kinds of difficulties might befall us, and what forms of preventative maintenance we must do to keep our warranty in effect.

One of the commonest problems, especially for smallish terrestrial planets, is the loss of essential atmospheric gases. Usually this is a slow, predictable process. Small planets, like Mars, have low enough escape velocities so that light volatiles, especially hydrogen and helium, can readily escape. Loss of the inert gas helium is of no biological significance, but loss of hydrogen is fundamental. The principal source of hydrogen gas on terrestrial planets is UV photolysis of water vapor. If the star is hotter than the Sun, its spectrum will contain not only more total radiant energy, but also a higher proportion of ultraviolet radiation. The rate of production of hydrogen will be high, and the loss of that hydrogen will be efficient. These planets will tend to dry out over the course of billions of years.

In the presence of extreme ultraviolet light, the dissociative recombination process can also strip carbon, nitrogen, and oxygen from bodies up to about the size of Mars. Large terrestrial planets are immune to this process.

Planets with more surface area per unit mass than Earth (smaller radius), also tend to lose their internal heat more efficiently. Crustal recycling either cannot get established (as on Mars), or begins but soon freezes up as the planet cools and its crust becomes thicker, cooler, and more rigid. Weathering reactions involving atmospheric gases then produce oxides, carbonates, sulfates, halides, and clay minerals, but cycling of these volatiles back into the atmosphere becomes almost impossible. The reactive gases in the atmosphere largely disappear irreversibly into the ground. On a body the size of Mars, there are at best one or two active volcanoes after several billion years of cooling. These volcanoes bury, and to some extent bake and outgas, the sediments produced by weathering. Volcanoes also transport volatiles from deep within the planet and spew them out into the atmosphere. But it is scarcely plausible that a single volcano's emissions, however large, can keep up with an entire planet's worth of gas removal by weathering. Also, as on Mars, low temperatures remove certain crucial atmospheric gases, especially water vapor and carbon dioxide, by precipitating them at the poles as snow.

Large terrestrial planets, the size of Earth or larger, in orbit around cool (G, K, or M), main sequence stars, need not fear water loss. And of course the enormously more massive, colder, and hydrogen-rich gas-giant planets and brown dwarfs are unaffected by hydrogen loss.

Explosive ejection of atmospheric gases by comet and asteroid impacts also preferentially affects planets of about the size of Mars or smaller. However, the population of bombarding bodies declines very rapidly early in planetary history, during the cleanup phase of planetary accretion. Therefore atmospheric loss is likely to occur at a significant rate only during the earliest days of the presence of simple life. Reduction of the atmospheric pressure by a factor of one hundred may occur early in planetary history for a Mars-size body, but the rate of atmospheric erosion several billion years later may amount to a loss rate of only 1 percent or so per billion years.

On planets on which life has already originated, large impacts can devastate the biosphere, resetting the evolutionary clock to zero — or perhaps to 1:00 A.M. This may happen on any terrestrial-type planet, no matter how large, but (with a few rare special exceptions) such impacts are almost certain to occur in the first few hundred million years of the planet's evolution, before life has had enough time to develop advanced, complex forms.

The motions of a planet may also change with time, often in ways that impact the habitability of its surface. The previous chapter dealt with ejection of planets from planetary systems into interstellar space, a process rare enough for habitable planets (which tend to be very close to their parent star and very strongly bound by gravitation) that we need not worry more about it.

The long-range gravitational interactions of planets with each other, especially with gas-giant planets, and the tidal influences of their moons and sun, can be very important. These forces drive the 120,000-year oscillation of the tilt of Mars's spin axis, periodically vary the eccentricity and inclination of Mars's orbit around the Sun, and cause the spin axis of the planet to precess. The eccentricity oscillates up and down with an amplitude of plus or minus 0.03 with a period of about 100,000 years, modulated by a slow periodic drift with a period of about 2 million years. The eccentricity variations do not change the mean temperature of the planet, but do strongly effect the annual range of temperatures. The minimum eccentricity can drop to zero, or range as high as 0.12, at which time annual temperature variations are dangerously extreme. The inclination of the orbit, which has no direct effect on climate, ranges from zero to about seven degrees.

The tilt (obliquity) of the spin axis can be periodically pumped up and down by these interactions. Mars, for example, has a spin-axis tilt that oscillates rapidly with a period of about 120,000 years, modulated by a slow cycle of about 1.2 million years' duration over which the range of variation of obliquity ranges from near zero to over twenty degrees. The obliquity can be as low as thirteen degrees, and, depending on which theory of the dynamics of Mars one adopts, as high as thirty-six to forty-two degrees at maximum. When we discussed Uranus, we mentioned that the axial tilt of Uranus, ninety-seven degrees, is so high that the average temperature of the poles actually ends up higher than that of the equator. This strange situation actually requires an axial tilt of at least forty-five to fifty degrees. Mars, therefore, can never quite wander into this state. Earth's axial tilt, by comparison, varies only between twenty-one and twenty-six degrees, moderated by the strong tidal interactions between Earth and the Moon that prevent external tidal forces of the Sun and planets from knocking Earth's axis around too severely.

The complex effects of precession, axial obliquity, and orbital eccentricity on the Martian climate were mentioned earlier; now we must realize that the axial tilt and orbital eccentricity are themselves variable in even more complex ways. Real long-range climate forecasting on Mars is a very difficulty business, requiring as much expertise in celestial mechanics as in meteorology.

In general, planets may be vulnerable to despinning. Both Mars and Earth have lost significant spin energy into tidal acceleration of their satellites. We have seen that planets that take refuge close to small main sequence stars to keep warm do so at the risk of their own spin. Capture, like that of the Galilean satellites, into 1:1 spin-orbit resonance, or, like Mercury, into higher resonances, is a common consequence of being too close to a massive body. A planet in a 1:1 resonance, as we saw in Chapter 12, can have only moderate temperature conditions at its terminator, the permanent sunrise/sunset line girdling the planet pole to pole. Every other location on the planet is either always too hot or always too cold. A planet with congenial temperatures in this "ribbon" of twilight tends to have so hot a subsolar point that water would boil away there, and so cold an antisolar point that the very atmosphere would condense and fall as snow of solid nitrogen and oxygen. Complete tidal despinning of a habitable planet will cause all of its volatiles to migrate into massive, frigid ice deposits covering the hemisphere of permanent night. Modest orbital eccentricity complicates matters by making the twilight zone oscillate back and forth, perhaps leaving the spin poles the only places without excessive temperature changes. But only rather large eccentricities would expose the nightside ice deposits to occasional solar heating. Somewhat larger eccentricities would permit higher-order spin-orbit resonances, such as 3:2, 2:1, or 5:2, which abolish permanent low-latitude ice altogether.

Strong tidal forces do not always mean a rotational lock on the central sun. One could imagine a system in which a planet and a massive satellite come into a perfect lock, with the orbital period of the moon exactly equal to the spin period of the satellite (like our Moon and Earth), and also exactly equal to the rotation period of the planet (unlike Earth and the Moon, but like Pluto and Charon). Then the planet can be protected against rapid despinning. All it need fear then is the loss of its satellite due to tidal interactions with its sun.

For moons that are somewhat smaller relative to their planets, the moon is not massive enough to soak up all the spin momentum of the planet at any finite orbital distance from the planet. The tidal interaction of the two bodies will continue to push the satellite outward to ever more distant orbits, until eventually the satellite can escape from its partially despun planet into an independent orbit around their sun. The problem with this scenario is that it leaves the planet and its former moon with intersecting orbits. Future close approaches are inevitable, and a devastating collision is possible, and even probable. If our Moon eventually escapes from Earth, it would pose just such a collision hazard.

Although tidal forces often cause satellites to evolve outward, this is not the only possible case. Imagine a satellite like the Moon, raising a tidal bulge in Earth's oceans and crust. Earth's rotation is much faster than the Moon's orbital motion around Earth (one day versus twenty-nine), so the tidal bulge on Earth is dragged a little ahead of the Moon's position by Earth's rotation. The gravitational pull of this bulge on the Moon tends to pull it a little forward; that is, to accelerate its motion about the Earth, transfer some of Earth's spin energy to the Moon, and lift it to an even higher orbit. The tidal interaction will then be a little weaker because of the increased Earth-Moon distance and because of the slightly slower rotation of Earth, but the force will still be directed so as to accelerate the Moon further.

But what if the Moon were so close that it orbited Earth in less than twenty-four hours? Then the Moon would speed on ahead of the tidal bulge it raises, and the gravitational attraction of the bulge on the Moon would tend to retard the Moon's orbital motion. Likewise, the attraction of the Moon's gravity would tend to accelerate Earth's tidal bulge and pull it closer to the Moon's position. There would be a constant loss of orbital energy by the Moon, and a constant acceleration of the spin of Earth. The Moon would approach Earth more closely, and therefore experience ever stronger tidal interactions. The Moon would accelerate its inward spiraling until it either crashed on Earth's surface or drifted inside the Roche limit, where it would be torn apart by the intense tidal forces. Assuming that the Moon is torn apart, it would form a rubble ring that, through collisions, would spread both outward and inward in the orbital plane. The inner edge of the ring would eventually reach the upper atmosphere of the planet, and debris would begin to rain down along the equator.

Because close satellites evolve rapidly and accelerate in their own rate of inward orbital evolution, such bodies are likely to have very short lifetimes. Their crash may occur during the era of final accretion of the planets, which would hopelessly obscure any possible geological evidence of the event. Satellite crashes would tend to occur so early in history that there is no reasonable possibility of advanced life forms being present.

As traumatic as an inward-spiraling falling moon would be, there are even worse possibilities. One of these is the encounter of a planet with a large escaped former moon in independent orbit as described earlier. Only very massive moons can behave in this manner, escapees from moon-planet pairs that we would call twin planets. Such planets are probably rare, but we really don't know. The most disheartening aspect of this scenario is that it has a very long fuse: Pandora's box may remain shut for billions of years of successful biological evolution, only to spring open before the appalled eyes of intelligent witnesses. The bright side of the matter is that the time that elapses between escape of the moon and its eventual catastrophic collision with the planet is probably at least one hundred thousand years, and possibly millions of years. The residents of the planet would have the strongest possible incentive to master space technology on a very large scale. The choice before them is threefold: fight, flight, or annihilation.

Another possibility for giant impacts is the collision of two sizable planetary bodies during the final stage of accretion, much as current theory imagines the origin of the Moon by impact of a Mars-size planetary body with Earth. Such an event would occur too early to be of much biological significance. A third possibility, much harder to assess, arises from the unsettling discovery that there are chaotic factors in the evolution of planetary orbits that not only make accurate predictions over tens of millions of years unreliable, but makes the long-term evolution of planetary orbits intrinsically and unavoidably unpredictable. A giant comet from the Kuiper belt may pass by Neptune and be nudged into a Jupiter-crossing orbit, and then be kicked by Jupiter's powerful gravity into an Earth-crossing orbit. A minuscule change in location or timing of the original orbit could aim the same body so close to Jupiter that it would be immediately ejected from the Solar System, hit Jupiter, or be fired into the Sun. We not only don't know whether such an event will happen in the next thousand years, we *can't* know. But we can see that no such event has occurred on any of the terrestrial planets in the last three to four billion years. Even though chaos can be lethal, it hasn't done us in yet.

But let's assume that law and order prevail in our stellar system — a place for every planet, and every planet in its place. Either the motions of the planets obey a higher law, or the local intelligent race has truly awesome powers of enforcement. But even under these apparently ideal circumstances, something can still go wrong. And this time the fault, dear Brutus, lies not in ourselves, but in our stars.

We have already encountered three different ways in which an inconstant star can wreak havoc upon its offspring. First, there is the slow luminosity increase experienced by main sequence stars over the course of their hydrogen-burning careers. A planet may be held in a deep freeze for billions of years, only to emerge into Edenic conditions too late for life to proliferate. Or, alternatively, a planet and its burden of life may be slowly roasted as its star increases in luminosity. Some significant percentage of Earthlike planets must fall victim to this process; however, life is both adaptable and robust. The biosphere may adjust the composition of the planetary atmosphere so as to fine-tune the strength of the greenhouse effect and moderate the temperature changes. Also, life may adapt to slow warming through natural selection. Something like both of these processes seems to have happened on Earth.

A second problem caused by the behavior of stars is flare activity. Flares are proportionately more important in the coolest, longest-lived main sequence stars. They may be devastatingly important only for stars so cool that their warm planets are very close and tidally locked, or in which the life zone is inside the star's Roche limit.

Third, every star eventually grows tired of humdrum life on the main sequence. Some simply slide away into oblivion, but these tend to be the ones with so little mass that they have enormous lifetimes. Others, in their midlife crises, develop a taste for flashy and spendthrift living, start lavishly burning their accumulated hoards of helium, and become giants — briefly highly visible, their lifestyles widely reported — until nature catches up with them! As we have seen, there is only one class of habitable body that might profit from entry into the giant phase, and that only for a time: a Europa in orbit around an appropriately distant star may have its ice layer melted. Its residents would then have an opportunity to learn astronomy, and possibly aerospace engineering, before conditions get worse. The presence of advanced beings implies (we guess) a billion years or more of biological evolution, which requires a parent star with a main sequence lifetime of at least a billion years, which in turn rules out a star massive enough for a supernova explosion. The specific combination of events required to bring about this scenario is improbable, but by no means impossible.

If one's own planetary system is safe and well policed, and one's star is reliable, there are still things that can go wrong. Other stars, often living by their own rules, may prove very disruptive. Close double-star systems with mass transfer can be very unpleasant neighbors, and distant companion stars in highly eccentric orbits can be disruptive near the time of perihelion passage, both by reason of strong gravitational forces and brief, powerful heating episodes.

Close encounters between independent stellar systems may also be disruptive, especially for Population II stars, which live in very crowded surroundings. The fate of such stars in dense, busy globular clusters, where close approaches are much more common than in the relatively sparse spiral arms, was discussed in Chapter 14. Close encounters between Population I spiral-arm stars are rare enough, and usually distant enough, so that we need not worry about them (unless, of course, we happen to be extremely unlucky).

But local Population I stars have another problem: about twice in each revolution around the center of the Galaxy, every Population I star passes through the heart of a spiral arm. There the density of stars is higher and the chance of encountering a star-forming region is maximized. Star-forming regions sound scenic and benign. Many textbooks refer avuncularly to "stellar nurseries." A more accurate depiction might be "tyrannosaurus hatcheries." These regions are infested with massive, dense interstellar clouds, all ripe for collapse. A planetary system passing through a giant molecular cloud complex can be stripped of gases and small bodies, bombarded by dust and gases, and even suffer dimming of its star due to huge concentrations of fine interstellar dust.

Sprinkled among these clouds are thousands to millions of young stars of all possible masses, including highly unstable titans with over one hundred times the mass of the Sun. T-Tauri stars in the ensemble are blowing off intense, hot polar jets of plasma. Brilliant new O, B, and A stars fill the region with coruscating ultraviolet and violet light, ripping apart molecules and making the more tenuous parts of the cloud fluoresce under the deadly radiation. And all the new stars with masses greater than about nine Suns will dash through their main sequence lifetimes, erupt as giants and supergiants, and then explode as supernovas in the midst of the "nursery." Such a nursery is no place for children — or other living things.

Our Solar System has been exposed to just such an environment roughly thirty times since the birth of the Sun and planets. It is entirely possible that some of the great extinction events in Earth's history were caused by passage through a star-forming region, perhaps at a time when a young, massive nearby star blew itself up. However bad it sounds, our ancestors survived.

IF WE were called upon to design a new planetary system to optimize the chances for the origin and development of life, what design features would we incorporate, and which would we avoid?

First, we would prefer a single star. Very close low-mass doubles or very distant multiples that never get too close to each other would be just about as good. We require that the star be on the main sequence. Spectral classes O, B, A, and most (or all) of F class have such short lifetimes that we would not choose them as a place for life to evolve to great heights. (But a mature starfaring race with ecological sensitivity might prefer colonizing F systems, because doing so would minimize their impact on native life forms.) The coolest, faintest M stars are unpromising because any planet close enough to its primary to melt ice would be tidally locked or even torn apart by tidal forces.

Population II stars are not attractive places to build planets because of their paucity of heavy elements: you can't build a terrestrial planet or an icy satellite out of nothing.

For terrestrial-type planets, a mass similar to Mars would suffice to make the presence of liquid water and the origin of life possible. But a substantially larger mass, perhaps 80 percent of the mass of Earth, may be required to keep the planet geologically active, with crustal recycling. The largest tenable Earthlike planets may be ten or more times the mass of Earth, but such bodies would be in danger of capturing solar nebula gas and becoming Uranian or Jovian planets. If they avoided such a fate, then they would be ocean planets with little or no land area and high surface gravities. They would be unsuited to human origin and evolution, but might be fine for intelligent aquatic beings and their food chain.

For Europas, a somewhat broader range of star types can be allowed. There is even the possibility of a Europa evolving into a water planet during the red giant phase of its sun. Europas can exist around both giant planets and brown dwarfs. They are the most portable bodies that can support biospheres; in some cases, when sub-ice melting is maintained by tidal interactions, they are actually capable of surviving travel through interstellar space without the presence of a main sequence star. This prospect lends Europas a peculiar significance and interest. It makes the prospect of exploring Jupiter's Galilean satellites exceptionally interesting.

Of course, much of this discussion is predicated upon life elsewhere being similar to life as we know it. It is an exciting and challenging game to try to imagine distinctly non-Earthlike life that is well adapted to equally non-Earthlike planets. This topic, though not susceptible to any absolute conclusions, merits more careful consideration.

17

THE PLURALITY OF HABITABLE WORLDS

SO MUCH DEPENDS ON WHAT we think is essential for life. Thus far we have assumed, almost without argument, that life everywhere is based on the chemistry of carbon and on liquid water as a solvent; that is, that life always means life as we know it here on Earth. We have limited our search for habitable worlds to those that have liquid water somewhere on or in them. But the Universe is a big place. Even if we are right about the rule, there are probably exceptions to it.

Now is the time to ask about life as we don't know it. What other solvents might be available to replace water? What other molecular backbone may accommodate the enormous complexity of a genetic code? What alien environments beyond those we have already approved might possibly harbor such alien solvents and biochemistries? As we ask these questions we labor under a heavy conceptual burden: we have only one example of a biochemistry, and envisioning anything other than modest variations on the familiar is a mind-taxing effort.

On a cruder, purely morphological level, the tendency of science fiction writers, artists, and movie producers to imagine aliens as variations on familiar terrestrial architectures is well known. As the Lectroids from Planet Ten pointed out in the film *The Adventures of Buckaroo Banzai*, we are "monkey boys." Alien intelligence suggests to us something bilaterally symmetric, with two arms and two legs, two eyes, two ears, and a digestive plumbing system aligned along the symmetry plane. By our standards, a grotesque alien intelligence is one that has almost all of these traits, but one or two of them are changed to convey a feeling of strangeness: they are given a second set of arms, or three fingers, or green skin, or enormous canine teeth, or a third eye. They are grotesque because we interpret them as mutated or flawed humans. And what of benign alien intelligences in fiction? They also partake of the same general set of human architectural features, but they have big blue eyes or cute ears, are furry and cuddly, often much smaller than people, and often have oversized heads (like human babies). More inventive imaginations pattern alien intelligences after dolphins or slugs, spiders or starfish, usually to create specific emotional connections. Most inventive, and least resonant with the psyches of the audience, are living alien intelligences that you would not readily recognize as intelligent, or even as life forms. Slime, rocks, and amorphous energy fields come to mind.

If our imagination is so terribly limited with respect to the outward physical appearances of aliens, how can we possibly hope to see inside them and guess what biology, physics, and chemistry makes them tick? Isn't this completely futile? I don't think so. On the level of the principles and materials governing structure and chemical function, the laws of chemistry and physics are universal. Outward morphology, on the other hand, is strongly shaped by a bewildering variety of possible adaptations to wildly diverse alien environments, as well as to chance mutations. I think we can do a better job with alien chemistry than with alien biology. Chemistry is a lot simpler.

LET'S START with the solvent. First, why do we need a solvent? Because chemical reactions in general do not take place inside solids: they take place on surfaces, in solutions, and in gases. Surfaces are fine locations for molecules and atoms to sit *in order to react with something else*, such as gases or solutions. Gases are great places for rapid reactions. Anything that readily forms a gas can be used. But at normal temperatures (from absolute zero to about 500 K) the overwhelming majority of all substances is either something other than gases, or the substances decompose into simple molecules when they become gases. Above about 500 K, most substances decompose when they vaporize. Complex organic materials are for the most part intolerant of temperatures much above 373 K (the normal boiling point of water at Earth's surface). We call the process of heating organic material to higher temperatures "frying." Such temperatures denature proteins, pyrolyze sugars, and melt fats. Somewhat higher temperatures irreversibly destroy all organic matter and virtually all other complex molecules. Further, in gases the density is typically a thousand times lower than that of a liquid solution. Collisions between molecules, which are essential to making reactions possible, are millions to billions of times less frequent in a given volume of gas than in the same volume of liquid. The complex products that are formed will generally be much less volatile than the reactants that make them: they will tend to condense and be unavailable for further reaction. Gases are a very poor medium for the chemistry of life.

Some liquids have the ability to dissolve a wide variety of complex materials, making vast numbers of their molecules available for reaction with each other in a small volume. The ideal solvent has the ability to dissolve a wide range of other materials without destroying them. Of all the tens of thousands of liquid solvents known to chemistry, there is simply no question which one is the best solvent. It is liquid water. Water's versatility arises from its polar nature: it has an electronegative end (oxygen) and an electropositive end (the hydrogen atoms) that can attach to a vast variety of other molecules. As if these were not reasons enough for accepting the predominance of water, consider this: water is the most abundant chemical compound, and by far the most abundant liquid, in the Universe. It not only works best; it's available in vast amounts.

Given this perspective, we should ask whether any other polar molecule is anywhere near as abundant as water and anywhere near as good a solvent. There are, in fact, very few viable candidates.

A number of different polar molecules are available in nature. Second after water is ammonia. Ammonia is a little less polar than water, but is still a very good solvent. It is also abundant in the cosmos, since nitrogen is one of the five most abundant elements, after hydrogen, helium, oxygen, and carbon. Ammonia is stable at low temperatures and high pressures, especially in the presence of hydrogen gas. It is a known constituent of the atmospheres of the giant planets and comets, and a strongly suspected major constituent of cometary and icy-satellite ices. But liquid ammonia is a rather different story. The liquid range of ammonia at normal pressures is from about −78 Celsius to −33 Celsius. It therefore boils (at one atmosphere pressure) at a temperature thirty-three degrees below the freezing point of water ice. It sounds like the ideal low-temperature solvent.

The problem is that ammonia can be present in the liquid state only under conditions in which water ice is already present. The combination of ammonia plus water ice allows the formation of a water-ammonia solution at temperatures as low as −100 Celsius. In general, almost anywhere in nature that ammonia might be within its liquid range, it actually forms a solution in water instead. In effect, ammonia is a widely available antifreeze that makes water available as a solvent down to temperatures far below the normal freezing temperature of water. Ammonia is indeed the key to fascinating chemistry, but only because of its collaboration with water. The ammonia-water solution is a better solvent than liquid ammonia, and has a vastly wider temperature range at normal pressures (from −100°C to +100°C, instead of only −78 to −33°C for pure liquid ammonia). Elevated pressures extend both temperature ranges somewhat, but the water-ammonia solution is always the winner. It nearly doubles the liquid range of water — and quadruples the liquid range of ammonia.

What about other polar molecules? There is carbon monoxide, which melts at −207° Celsius and boils at −190°C. At these very low temperatures, CO, which is moderately polar, actually dissolves very little. The narrow liquid range means that it is very hard to make and preserve liquid CO. In fact, its presence has never been demonstrated, or even seriously suggested, anywhere in the Solar System.

Liquid silicates, which can be made on many Solar System bodies, are polar and have interesting chain structures, but at the temperatures necessary to melt rock virtually all complex molecules, both organic and inorganic, are unstable and dissociate into simpler fragments. Sulfur, the tenth most abundant element (after H, He, O, C, N, Ne, Fe, Si, and Mg) can form liquid sulfur dioxide (liquid range −75 to −10°C) or hydrogen sulfide (−83 to −62°C). Sulfur dioxide is not nearly as universal a solvent as water, but it is pretty good. Further, there are some settings in which liquid sulfur dioxide is a plausible natural material — for example, beneath the surface of Io. The narrow liquid range of hydrogen sulfide and its tendency to form solid sulfides makes it a less credible solvent. Further, in nature hydrogen sulfide condenses only under circumstances in which the much more abundant water ice and ammonia are both condensed. The hydrogen sulfide would simply react with ammonia to make ammonium hydrosulfide and dissolve in the cold water-ammonia solution. If we were to devise a method of separating hydrogen sulfide from ammonia and water, its slight polarity, its very narrow liquid range, and its low abundance relative to water would all render it a far less useful solvent.

Hydrogen chloride and hydrogen fluoride are both polar molecules, but both are very much less abundant than water. They both react readily with a wide range of minerals to make solid salts. Liquid hydrogen chloride dissolves readily to make a dilute solution in the vastly more abundant water. Hydrogen fluoride is even less abundant and even more reactive toward minerals. Neither one seems a plausible candidate.

Nitrogen oxides are also polar liquids, but they are of such low stability that they are not found in high concentrations anywhere. Again, the liquids have never been found to occur in nature, and both are extremely soluble in the much more abundant water. These also seem to be poor candidates.

The preferred solvents then appear to be water-ammonia solution (more or less pure water toward the higher temperatures) and conceivably liquid sulfur dioxide. Every other substance is either far less abundant in nature or a far inferior solvent, or both. Among the other possible abundant fluids, liquid methane, nitrogen, neon, and argon have no dipole moments, and are poor solvents for anything except each other and simple hydrocarbons.

Now we can turn our attention to the actual structural material out of which the molecules of life and life's blueprints can be made. The guiding principles in choosing potential building materials for life are that it must be possible to use them to construct large, information-rich molecules that permit coding a large amount of information; that these molecules form long chains that can be "read" and assembled sequentially; that these molecules be neither so stable that they cannot react nor so reactive that they spontaneously fall apart; that a plausible solvent be available to mobilize them without completely destroying them; and that they be made of materials that are reasonably abundant in natural astronomical and geochemical settings.

A number of substances permits the construction of long chains. The best known is, of course, carbon. Very long carbon chains with attached hydrogen atoms (hydrocarbons), however, are more characteristic of crude oil than of living organisms. Where complex hydrocarbons are made by life, they are generally end products, usually made of large numbers of identical units, such as the terpene family familiar from the odor of evergreens and turpentine: they are not involved in further reaction cycles or in coding genetic information. In a sense, it is not really true that carbon forms the backbones of biological molecules. Instead, it is carbon's compounds with hydrogen, oxygen, phosphorus, and nitrogen that provide the building blocks out of which long chains are assembled. Proteins have C, O, and N atoms in their backbones; DNA and RNA are held together by chains of alternating phosphate and sugar units in which the "molecular backbone" is dominated by oxygen and phosphorus. Chains of alternating phosphate and sugar units provide a long, stable structure for carrying a sequence of organic bases, one of which is attached to each sugar. There are only four organic bases used in DNA, simple molecules of hydrogen, carbon, nitrogen, and oxygen called guanine (G), cytosine (C), adenine (A), and thymine (T). They spell out long messages such as CCGATCGGGAATGTCTAAGT, which carry a rich store of information in the form of a long ribbon-like molecule that can be read and copied.

Phosphates can not only link to form chains in which phosphates alternate with sugar molecules, but also form chains in which phosphate units link up directly with each other to form chains called polyphosphates. Many phosphates dissolve readily in water, although some, such as magnesium, calcium, and ferrous phosphate, are very insoluble minerals. In normal solutions of phosphates in water, the phosphates exist as separate ions. They link up to form long chains only in the most concentrated solutions, effectively in concentrated phosphoric acid. These chains are reactive, constantly breaking and reforming. Further, they are not a rich medium for coding complex information. The alphabet is simply too limited. Let us represent each phosphate unit, PO_4^{3-}, as a letter p. Then the chains of phosphates spell out pp, ppp, pppp, ppppp, and so on. These chains constantly break and rearrange on their own, so even having a convention that pppp means one thing and ppppp means another, the sense of any message is lost in microseconds due to spontaneous rearrangements. Also, the requirement that the solvent be concentrated phosphoric acid is a troubling one. In an oxidized environment, in which phosphorus forms bonds with oxygen (to make phosphates) instead of with hydrogen (to make phosphine, PH_3), the phosphate is rather efficiently bound up as insoluble phosphate minerals. We have no known examples of such highly concentrated phosphate solutions occurring naturally on Earth or on any other body. Low-concentration solutions of phosphates are possible, and they could in principle be stabilized by reacting with something that hinders their ability to react with minerals or metal ions — such as sugars. But that is just another way of saying that the phosphates don't link together to form long chains in nature unless a carbon-based organic chemistry is present.

Note that, in chemically reduced (unoxidized) systems, phosphates are not stable. The phosphorus hydrides have a very limited ability to link up in chains: the practical limit is the unstable molecule diphosphine, with two phosphorus atoms.

Certain other less common elements are noted for their ability to form chains. By far the most successful are sulfur and iodine.

Sulfur at low temperatures usually consists of stable puckered rings, each containing exactly eight sulfur atoms. Ordinary yellow sulfur is just crystalline S_8. Heating sulfur well above its melting temperature of about 120 Celsius breaks open many of the sulfur rings to form sulfur chains, which are very reactive. The chains readily add and lose single S atoms or short chain fragments consisting of S-S, S-S-S, and so on. The ideal solvent for these chains is liquid sulfur itself. The problem of coding information on short, highly unstable chains of identical units is the same for sulfur as it was for polyphosphates. Incidentally, crystalline eight-atom rings of sulfur dissolve readily in carbon disulfide, but the long, irregular chains of sulfur atoms do not. Although liquid sulfur occurs naturally on Earth and Io, and almost certainly was once present on Mars and Venus as well, and sulfur is a common element on all solid planets, liquid sulfur is generally rare. It is interesting that the short S-S chain has been successfully incorporated into terrestrial biochemistry as part of the structure of the essential amino acid cystine. But longer sulfur chains, even when present, are transient, unstable structures.

Sulfur forms not only neutral chains of S atoms, but also polysulfide ions, with the general formula $S_x^{=}$. Highly concentrated solutions of sulfide chains can be made by dissolving sulfur in a concentrated solution of sodium or potassium sulfide in water. Chains of a dozen or so atoms exist in solution, but the chains are extremely reactive and fragile. Crystalline solids with up to four or five sulfur atoms in the polysulfide chain have been prepared by chemists. Polysulfides seem no more promising as the building blocks of life than neutral sulfur chains.

Under some circumstances, the halogens can also form chains. For example, chlorine gas dissolves readily in concentrated solutions of chlorides such as hydrochloric acid or sodium chloride to make the trichloride ion, Cl_3^{-1}. But this ion readily rearranges, and decomposes at lower solution concentrations. Bromine behaves similarly, but a little more readily. Of course, bromine is a rare element compared with chlorine, and has never been observed in nature to reach high enough concentrations to make tribromides stable. Iodine, the next heavier (and even rarer) halogen, continues the trend with a vengeance. Concentrated solutions of iodides in water readily dissolve large quantities of solid iodine to make beautiful deep purple solutions. Many of the polyiodides can be crystallized as purple, almost metallic-looking crystals. The larger the size of the positive ion in the compound, the larger the negative iodide chain that can be accommodated by the crystal. Potassium iodide readily forms potassium triiodide, KI_3, and several crystalline pentaiodides have been made. In solution, chains of nine and even eleven iodines have been identified. But all of these chains are extremely fragile and subject to spontaneous breaking and recombination. Again, there seems to be no way to code information using polyhalides.

Finally we come to silicon. Silicon offers three opportunities for forming complex molecules: silicon-silicon bonds, silicate linkages, and silicones. Silicon-silicon bonds are the least promising of the three, being limited to about three silicon atoms in a row. But the other silicon-bearing structures are more interesting.

Silicates are astonishingly versatile in their ability to form one-, two-, and three-dimensional structures. Silicon atoms, whenever possible, surround themselves with four oxygen atoms in the shape of a tetrahedron, with the silicon atom hidden in the center. This structural unit is the silicate ion (SiO_4^{-4}). A few silicates, such as those of sodium and potassium, dissolve in water, but most are very insoluble minerals. At high concentrations, silicate ions can link together by sharing oxygen atoms between two different tetrahedrons, making disilicate and tetrasilicate ions. But the full expression of the ability of silicates to link together is found only in solvent-free systems such as molten rock and solid silicate minerals. Some minerals contain very long chains of silicate tetrahedrons. Others, such as the mica family, contain enormous two-dimensional sheets of linked silicate tetrahedrons. Others contain three-dimensional networks of silicates. The variety of structures is amazing. But such structures are hidden in the frozen interiors of mineral grains, where they can be neither read nor used chemically. It is true that complex silicate structures can also be found in molten silicates, but at the temperatures of liquid rock, rearrangement of these structures is very rapid. There are no low-temperature solvents in which these structures form.

There is a way to stabilize silicon-bearing chains at low temperatures: by using the silicone structure. The backbone of the silicones is a chain of alternating silicon and oxygen atoms. Each silicon also carries two attached atoms or groups of atoms such as H, Cl, F, CH_3, and so on. Substituents such as Cl and F are readily removed by water, leaving hydroxyl (OH) in their place. A unit in the chain therefore might be -O-SiF_2-, repeated endlessly. If water is present, the possible units will be -O-SiH_2-, -O-SiHOH-, and -O-$Si(OH)_2$-, which we can denote a, b, and c. These "letters" can then be linked together in a chain to make "words" such as aaaa, aabaaab, cbaabaccbaabbc, and so on. If carbon-based side chains are possible, then a vast number of different silicone letters could exist, from -O-$SiHCH_3$- and -O-$Si(CH_3)_2$- to lengthy hydrocarbon or other organic chains. In other words, silicon chemistry is most promising if ordinary organic chemistry is available. But, if carbon-based chemistry is available, there is no need for the silicone backbone! We may fairly count this as strike one against silicones. The absence of suitable solvents for silicones, and the ready availability of water and water-ammonia solutions for carbon-based organics, constitute a second strike against silicone-based life. But we cannot quite call the idea out on strikes. The complete absence of silicones in nature leads us to anticipate a K on the score card, but certainty is not so easy to come by. We should also note that, observationally, silicone-based life on Earth is limited to Las Vegas and a few other similar ecosystems, where it seems to confer no reproductive advantage.

But there are even broader options open. We may ask, "How common are planets that are habitable by life forms with very different chemistry from Earth life?" From our position of almost perfect ignorance, we can do little better than suggest that liquid sulfur dioxide on a grossly Io-like body is a distantly conceivable medium for complex carbon-based chemistry, or that silicone-based chemistry, in the presence of an as-yet-unknown appropriate solvent, might also be possible. But the circumstances necessary for making a water-free sulfur dioxide world are demanding and probably rare, and the conditions needed to generate a silicone-based chemistry are so bizarre as to be virtually incredible.

Despite our best efforts to step aside from terrestrial chauvinism and to seek out other solvents and structural chemistries for life, we are forced to conclude that water is the best of all possible solvents, and carbon compounds are apparently the best of all possible carriers of complex information. We have not proved that all others are impossible, nor that Earth is the best of all possible worlds. Plotinus's question remains largely unanswered.

THE QUESTION asked since ancient times is, "How common are Earthlike worlds?" This question can be answered only if we specify how Earthlike such worlds need to be to harbor life of their own. The most specific, almost ludicrous, interpretation is that Earthlike worlds are those where we could walk around, breathe the air, drink the water, and eat the local foods without discomfort or danger. Such worlds must be very rare, and probably do not exist: even on Earth, this description fits Eden better than any real state or nation. What about worlds where we could breathe the air, drink the water, and raise foods brought with us from Earth? They must be much more common. Alternatively, is it necessary for a habitable world to have oceans and oxygen? What if there is almost no gaseous oxygen? Do we lose interest? After all, for most of the history of Earth, there was abundant life and no free oxygen! And what if there is a multitude of small, isolated seas and lakes rather than global oceans? Or if the seas or oceans were mostly iced over? Is a planet Earthlike if it has a year ten Earth days long, or rotates seven and a half times per year, or orbits a gas-giant planet which in turn orbits a star? Is it Earthlike if it abounds in liquid water but has no sun in its sky, only a warm, feebly glowing brown dwarf?

The more we look over the vast range of possibilities for habitats of life, the less we are inclined to ask — or try to answer — this question in its narrowest form. We would be better advised to broaden the question to, "How common are planets that are habitable by life that is chemically similar to Earth life?" Here we assume chemistry based only on compounds of carbon, oxygen, hydrogen, and nitrogen. The interesting answer is that there may be a wild variety of such habitable worlds, some of them clearly un-Earthly in a variety of ways, but still having liquid water, organic matter, and energy sources for powering the origin and maintenance of life. Some are Earths and some are Earthlets, some are Earthissimos and some are Europas, some are Earthlike moons of giant planets and some are sunless worlds with temperatures moderated by a brown dwarf and chemistry driven by tidally-powered hydrothermal activity. To each according to his tastes. But each will have its own favorite variations on that basic organic chemistry, adapted both to local conditions at the time of the origin of life and to the nature of the continuing source of chemical energy at the bottom of its food chain. Life adapted to such disparate conditions must be endlessly diverse, in keeping with the expectations of Galileo, Bessel, and others. Who are we to say which is the best of all possible worlds?

If human explorers should someday encounter a planet inhabited by native life forms, could we make ourselves at home there? The best answer we can offer is that it is highly probable that the chemistry of those life forms (not necessarily its biological architecture) is comprehensible and based on carbon compounds. But this is no guarantee of the presence of mobile beings with high rates of energy expenditure (those which, if they lived on Earth, we would call animals). The presence of alien animals is in turn no guarantee of high intelligence, and the presence of high intelligence is no guarantee of easy communication. But such planets, whether or not inhabited by animals, are possible homes for our descendants. Those who travel to the stars will go equipped with not only the tools of physics, but also a deep understanding of the chemistry and genetics of life and the laws that govern the behavior of planets. They will be seeking not planets where humans would be instantly at home, but worlds that can be accommodated to a mankind that is itself willing and able to change. We need only a place that we can meet halfway.

18

WHERE DO WE GO FROM HERE?

TWO AND A HALF MILLENNIA ago the ancient Greeks had strongly held, well-developed opinions about the nature of the universe and the plurality of worlds. Some argued that our Sun was the only sun in the kosmos and that there were no other planets besides Earth; others, especially the atomists, argued that there was an infinite number of Suns and planets, with life everywhere. Neither conclusion was correct, and neither school should receive credit for pioneering the modern worldview. The reason they were unable to resolve their disagreements is simple: the various schools of thought derived their opinions by logic from a set of philosophical and religious premises that were in no way based on observation or constrained by reality. They neither based their explanations of the Universe on quantitative observations nor thought to test them in this manner. In short, they speculated shamelessly, concerned more about their standards of logic than whether their initial assumptions or final conclusions matched reality.

To a significant extent, whenever we in the modern world are forced to extrapolate our ideas into realms where observations are sparse and difficult, we are prone to fall into the same trap. The difference is that modern scientists do base their projections upon large bodies of other data that appear relevant. The astrophysicist specializing in stellar evolution who runs computer models of the lifetimes of stars, the observational astronomer who watches T-Tauri stars go through their antics, the planetary scientist who sends spacecraft to explore the other planets of our Solar System, the meteoriticist who examines the subtlest details of a bewildering diversity of meteorites in the laboratory — all have mastered large bodies of specialized knowledge, and all are generally aware of the limits of their expertise. The nonbiblical parable of the blind men and the elephant comes to mind: each has studied a relevant aspect of the problem of the plurality of worlds, and each has only a limited awareness of the observations and theories of the others. There is a tendency for the expert on magnetic fields to see such fields playing a dominant role in preplanetary stellar nebulae; for the chemist to see condensation and the kinetics of chemical reactions as by far the most important factor; for the fluid-mechanician to see turbulent transport of angular momentum as the key to the origin of planets. Like the blind men, curiously, they are all correct—up to the point where they assert the exclusivity of their own interpretations. Then they become simply partisan philosophers, out of their depth in a vast sea of unfamiliar data.

How can science make progress in such a setting? By the same techniques it has always used: each expert wandering half-blind into a new arena of research must construct hypotheses that are designed to explain all the data (practically, this means all the data familiar to the expert). Normally, in new areas of endeavor such as that discussed in this book, many specialists in several different sciences may propose quite different hypotheses. The next step, which is peculiar to the scientific method, is to subject these hypotheses to *quantitative* test. Science is certainly not equivalent to the body of factual knowledge that it has amassed, nor to the currently accepted body of theory: rather, science is a method of developing succinct and universal theories that accurately explain our observational data. Science not only contains a process of testing hypotheses against reality—science *is* the process of testing hypotheses against reality. The only ideas that need fear the scientific method are those that are demonstrably at odds with observation.

Astronomy is different from most other sciences in that we usually cannot carry out experiments on astronomical bodies. Meteorites have been an exception to this generalization ever since their extraterrestrial origin was first widely accepted early in the nineteenth century. The bodies of the solar system, including the Sun, have also been accessible for in situ experimentation, and even sample return, since the 1960s. But interstellar probes (even interstellar meteorites) are not yet available for our use. We must for now be content with predicting the results of passive astronomical observations that have not yet been carried out, including those that have not yet shown that they are technically feasible. One such observational technique is the astrometric method of detecting extrasolar planets, which, unlike the radial-velocity technique, has yet to bear fruit. There is no doubt that the technique works in principle and that it will soon be producing useful results. The rapid advance of telescope and detector technology, and ever-broadening spectral coverage from radio waves to gamma rays, give us better access to relevant data every year. Other techniques for finding and characterizing extrasolar planets should not be far behind. The data needed to test our hypotheses will not be long in coming.

What can we hope to learn about the planetary systems of other stars from foreseeable Earth-based observations over the next ten to twenty years? First of all, we will be able to correlate the results of astrometric and radial-velocity measurements on many specific stellar systems. These two indirect techniques provide complementary, not identical, information.

The second step will be to develop techniques for direct detection of other planetary systems, to actually "see" and perform crude spectroscopic measurements on giant planets and brown dwarfs orbiting about other stars. These techniques will almost certainly be applied first in the infrared part of the spectrum where most of the light and most of the useful spectral data are to be found. Preliminary results can be expected in the near future.

Third in order, and most difficult by far, is the detection of terrestrial planets orbiting other stars. Telescopes, probably interferometers, in the fifty- to one-hundred-meter diameter class would have to be used for this search, preferably based far out in space, and certainly free of the interference of Earth's atmosphere. Meter-size individual mirrors placed fifty to one hundred meters apart on a very rigid framework would suffice to detect an "Earth" in orbit at 1 AU from a G2 star at distances of ten to twenty parsecs (2 to 4 million AU) from the Sun, a volume that contains up to ten thousand stellar systems. Even the faint zodiacal dust cloud associated with the asteroid belt would be a potential source of interference, requiring that the observatory be placed at least beyond the Hilda asteroids, and probably beyond the Greek and Trojan families on Jupiter's orbit, 5.2 AU from the Sun. Such interferometric detection can provide a low-resolution infrared spectrum of all the planets of a star lumped together. Since the largest planets are typically farthest from the central star, the total infrared brightness of several of the largest planets will be surprisingly similar to each other. Thus the discovery of many planets may be possible with a technique that simultaneously gives both a direct detection and some spectroscopic information.

Finally, we should mention the possibility that very large radio telescopes may detect natural emissions or even, if they exist, artificial broadcasts from other worlds. Very large radio interferometers in space, their individual receivers electronically linked through a computer, would make it possible to locate the sources of the signals very precisely. The Doppler shift of the frequency of the received radio waves would also (in the case of an artificial or natural maser radio source) permit measurement of the orbital and rotational properties of the source body.

JUST AS Earth-based observations of our own Solar System paved the way for interplanetary probes and manned expeditions to the Moon, so the discoveries of these space missions and the steady advance of technology have revitalized ground-based astronomy and driven many astronomical endeavors into space. A similar dynamic may apply to our discovery and study of extrasolar planets. Certainly space-based observational outposts are feasible and effective; indeed, we have seen that the ideal location for some of these devices is 5 AU or more from the Sun. But the nearest stars are about 200,000 AU away. Is there any prospect of sending unmanned scientific probes to visit other planetary systems? The answer, astonishingly, is yes — but with some very important caveats.

The most conservative approach to interstellar missions is to take advantage of existing chemical-rocket-propulsion systems. A multistage vehicle, after blasting off from Earth, achieves orbital speed and then escape velocity. The spacecraft launched in this manner may be targeted to pass close by Jupiter, Saturn, or even the Sun (after a close Jupiter swingby), firing its last-stage motor deep in the gravity well of a massive body where the impulse will do the most good. Such missions, like the *Voyager 1* and *Voyager 2* spacecraft, may achieve velocities of twenty-five to thirty kilometers per second relative to the Sun after they cross Jupiter's orbit outbound. They will escape from the Solar System completely, departing at a speed of about 5 AU per year. The time required for such a spacecraft to coast to a nearby star is therefore about forty thousand years. The most extreme types of missions accessible with chemical propulsion would involve diving close to the Sun behind a sun shade, then firing the last rocket stage at perihelion. Using this technique, departure velocities of about one hundred kilometers per second are possible. That would reduce the trip time to a nearby star to about ten thousand years.

Few humans can envision spending large amounts of time, energy, and money on a scientific research project that will not pay off for a hundred centuries. In fact, it is easy to think of reasons why this would be a silly thing to do. For instance, it fails completely to consider the natural course of scientific and engineering progress. Suppose a pilotless interstellar probe is launched in 2010 with a departure speed of 100 kilometers per second. If, over the ensuing century, it became possible to build a vehicle with a departure speed of 110 kilometers per second, then the latter probe would reach a target star 200,000 AU away nine hundred years before the one launched earlier! But the course of technological progress is far greater than 10 percent per century. A century ago the highest speeds ever attained by any human device were bullets and shells that traveled at about twice the speed of sound (0.2 kilometers per second). Within fifty years the record had improved to about 14 kilometers per second, a rate of improvement of 14,000 percent per century. The logic of launching interstellar probes is closely similar to that of buying a home computer: you always know that if you can wait about eighteen months there will be a new-generation machine with twice the speed, twice the memory, and twice the disk capacity of today's best machine — and the new one will cost 20 percent less! Logic dictates that you should never buy a computer — but practical necessity dictates that unless you buy something now you will never realize any advantages from this exponentiating technology. So it is with interstellar probes: at some point we must act. The question is simply put: When will we get started?

How can we return information from another star? Communication over interstellar distances is a moderate challenge, but if we are content with low rates of data transmission, we could build a workable interstellar probe this year. The principal difficulty faced by such a task is not communications, but component reliability over very long trip times. The entire design would have to be massively redundant, and the device must be able to diagnose malfunctions and reprogram itself to bypass failed components.

Interstellar probes have in fact already been launched. The two *Voyager* spacecraft are already about 100 AU from the Sun on their way into interstellar space. These spacecraft were designed to study the giant planets, not to make observations and send back data from another star. Nonetheless, they are departing on trajectories that will take them — *are* taking them — through interstellar space. If we had wanted to, we could have targeted them at other stars. As long as they continue to function, they can be tracked out to distances of at least a few percent of the way to the nearest star. But that limit is elastic: advances in radio telescope technology promise to increase the sensitivity of our receivers here on Earth at about the same rate that the signal from the *Voyagers* weakens due to increasing distance!

There is a clear incentive to shorten the travel time for interstellar voyages, both for the sake of the health of the components of the spacecraft and for the edification of the patient listeners who remain behind on Earth, hoping for some eventual gratification of their desire for information — and some return on their long-term investment. Shortening travel times dramatically (not just by tens of percent) requires that we transcend the limitations of chemical rocket propulsion. Within the last few decades, it has become clear that several types of propulsion systems that operate well beyond the performance limits of conventional rockets are possible. They all depend upon achieving exhaust velocities higher than chemical reactions can attain.

The speed of a rocket's exhaust increases with the square root of the temperature of the gas, and decreases with the square root of the molecular weight of the gas. The best (lightest) exhaust gas from a chemical rocket is water vapor, with a molecular weight of eighteen, made by the combustion of liquid hydrogen with liquid oxygen. If we could exhaust pure hydrogen, with a molecular weight of two, at the same temperature, it could outperform the hydrogen-oxygen rocket by a factor of three (the square root of 18/2). But, if we don't burn hydrogen with oxygen, how do we get the exhaust so hot? There are two principal options: use a nuclear reactor to heat the gas, or use a large inflatable mirror to focus many megawatts of sunlight upon a metal thrust chamber. These two types of engines are called nuclear thermal and solar thermal rockets, respectively. The advantage of these systems over chemical rockets is even greater if we can heat the hydrogen stream to higher temperatures than are achieved in a chemical flame. Such extreme temperatures require the most specialized, expensive high-temperature alloys, but can deliver up to nearly five times the exhaust velocity of a hydrogen-oxygen chemical rocket, nearly 20 kilometers per second instead of 4 kilometers per second. This advance alone would make it possible to achieve a speed increase of about 100 kilometers per second at perihelion, which translates into a cruising speed far from the Sun of about 330 kilometers per second (66 AU per year). The engine would then shut down and the payload would coast to its target. The trip time to a star 200,000-AU distant would then drop to three thousand years.

Even higher exhaust velocities can be achieved if we are willing to settle for lower thrust levels. We achieve the increase in performance by running the engine for a longer time. Several types of such high-performance engines have been studied in the laboratory, and some have even been used on a small scale in space. These engines use electrical energy to accelerate ionized gases to extremely high speeds. No thrust chamber is needed, and no exorbitantly expensive high-temperature metals are used. These electrical engines include the ion engine, the arc jet, and the plasma jet. They can deliver exhaust velocities of thirty to fifty kilometers per second. The biggest problem posed by such engines is the need for huge amounts of electric power, which can be supplied either by photovoltaic solar cells capturing sunlight, or by nuclear reactors. As with solar thermal rockets, solar electric rockets have the advantages of relatively light weight and freedom from radioactive materials. But the performance of such systems drops off rapidly the greater the distance from the Sun, since the intensity of sunlight drops off with the square of the distance from the Sun.

Solar sails, which work by using the pressure exerted by sunlight, also function well close to the Sun, but become impractically sluggish before reaching Jupiter's orbit.

Another type of electrical propulsion system involves firing metallic slugs out of an electromagnetic gun at extremely high speeds. Tests on Earth have demonstrated exit speeds of about nine kilometers per second, but space-based systems that can deliver speeds of 20 km/s or more should be possible. The electrical power must be supplied, as with the ion and plasma engines, by either the Sun or a nuclear reactor.

Enormously improved performance could be achieved by controlled fusion of light isotopes such as helium-3 and deuterium (heavy hydrogen). Essentially infinite supplies of these isotopes exist in the atmospheres of the giant planets. Schemes for retrieving these fuels from the atmosphere of Uranus are discussed in my book, *Mining the Sky*. A century from now we may be able to build a rocket that derives its power from the fusion of these gases, delivering an exhaust velocity as high as fifteen thousand kilometers per second. Ultimate speeds of as much as 3000

AU per year may be achieved. The time of travel to a star 200,000 AU away drops to less than seventy years. Thus the stars may actually be accessible on the time scale of a human lifetime.

Other options are opened up by the possibility of building very large solar-powered lasers in space, possibly inside Mercury's orbit. These lasers can be directed at solar-thermal or solar-electric spacecraft to enhance their performance far from the Sun, or they could be used directly to accelerate delicate thin-film solar sails to immense speeds. Such spacecraft, which do not need to carry any fuels or propulsion systems, would be blown through the interstellar voids by the pressure exerted by intense beams of carefully directed laser light.

It is, therefore, clearly technically feasible to send a pilotless probe to any of a large number of nearby stars. The trip times for these one-way voyages may be as short as a few decades. It is also technically feasible, although enormously difficult, to send a human crew to another star. It is by no means necessary for the crew to spend a lifetime playing solitaire during the seventy years of interstellar cruise. They may instead sleep away almost all of the trip in cold storage, taking turns being awake for a year or so at a time, each sleeping 90 or 95 percent of the time, and arriving at a physiological age only a few years older than they were at the time of departure. Astronauts could travel to a promising planetary system with Earthlike planets, or even travel there and back. Interstellar round trips to the nearest stars are possible within a human lifetime, using only known scientific principles — but the realization of such advanced propulsion systems requires the solution of a myriad of severe engineering problems. The job will not be easy, but it appears that it can be done.

CAN WE find Earthlike planets in orbit around nearby suns? Yes, if they are there. Are they there? The answer, as we have stressed many times in this book, depends on what we mean by Earthlike. Earth-size rocky planets, with or without oxygen-rich atmospheres, are probably common. Earth itself had an oxygen-poor atmosphere for most of its history. Indeed, even after life was well established in the oceans, the oxygen content remained low for billions of years. The right kinds of photosynthetic organisms, however, could pump out oxygen at such a rate that the air could be made breathable and an ozone layer established in short order. This happened to Earth naturally; in fact, the evidence of ancient banded iron formations suggests that oxygen may have risen many times before it became permanently established. Earth "terraformed" itself.

The word *terraforming* was coined a half century ago by science fiction author Jack Williamson. It implies human intervention in the regulatory cycles of a planet to make it more like Earth. There are many ways that this might be done. The favored approach depends to a very large extent on what is wrong with the planet.

It is possible, for example, that a planet with desirable composition is too warm for life to have arisen on it. Venus is a particularly nasty example of a planet that has become much too hot. The most direct cure for planetary sunstroke is to reduce the amount of light falling on the planet. Brightening up the planet, so as to reflect a higher proportion of the incident light, is almost never a viable option. Hot planets boil away the volatiles that might have condensed on their surfaces, and these substances (most notably water) condense in the atmosphere to form dense layers of brilliant, highly reflective precipitating clouds that lend the planet a high albedo (reflectivity). Even if one could find, transport, and inject a substance more reflective than these clouds, that material would not reside for long at the cloud tops, but would instead fall into the lower atmosphere and eventually end up on the surface. A far more effective strategy would be to reflect a large proportion of the incident light before it reaches the planet's atmosphere. In effect, we provide the planet with a sunscreen. Deprived of the intense heating that maintained high surface temperatures, the planet cools slowly by emission of infrared radiation until surface temperatures congenial for life are reached at the poles or on high-altitude landforms.

For planets that are a little too cold, such as Mars, we can envision decreasing the albedo by adding darkening matter. Low-albedo dust is common in the Solar System and in interstellar clouds, and is likely to be abundant everywhere. The darkening agent is most effective when applied to the brightest parts of the planetary surface, which is usually the polar caps. But such dark dust can be blown about by winds and buried. A single application would almost certainly give no more than a transient warming.

Another approach to warming the surface is to enhance the greenhouse effect by adding gases that, even in very small quantities, increase the infrared opacity of the atmosphere and help seal in the heat that actually reaches the surface. This is usually best done by releasing halocarbons into the atmosphere. Rather than transporting millions of tons of halocarbons over interstellar distances on the chance that they might prove useful, it would be far better to synthesize them on the surface of the planet where they are needed. Once the planet has warmed enough to shrink the polar caps and increase the water content of its atmosphere (and augment the greenhouse effect), the rate of injection of halocarbons may be reduced or even eliminated.

On many planets, a shortage of liquid water may be the most pressing problem. If the problem is that the water is frozen, then the warming scheme described above will suffice to break the ice. But if water is simply not there in any form, the problem becomes much more serious. Importation of water requires the availability of an appropriately rich, reasonably accessible source that does not lie deep in a gravitational well. Comets always come first to mind, but we must remind ourselves that comet impacts probably played a major role in stripping Mars of its initial atmosphere. Water-rich, relatively low-velocity asteroids are a far better solution. Ideally, the water should be supplied at such a low speed, or in such small chunks, that explosive blowoff of volatiles is unimportant.

Planets with insufficient atmosphere pose other problems. The atmospheric mass may be insufficient to provide protection against ultraviolet sunlight, meteoroids, and cosmic rays. Wholesale addition of carbon, nitrogen, and water can be accomplished by importing carbonaceous asteroids. But, in the early stages of colonization, the most useful solution is often to erect domes. Dome complexes can be extended indefinitely, building with local materials, until eventually the entire planet is roofed.

Planets with spin problems are yet another level more difficult. Resonant locks can be broken by large enough impacts; however, the planet will then tend to dissipate energy and drift back into the nearest accessible resonance. The game of playing with planetary spin requires an extravagant ante and allows no permanent fix. Once the spin state of a planet has been altered by a major glancing impact, the dust has settled, and a local population has grown, applying another nudge may require the temporary evacuation of the entire planet. A settled population would be very reluctant to call the spin doctor because of the level of disruption his house call would entail.

Most attractive are fixes that can be applied by biological intervention. Biological agents replicate and multiply themselves, and ecologies regulate both themselves and their planetary environment. We have seen that a planet like a Precambrian Earth may be best fixed by introduction of photosynthetic algae. Plants can fix the problem of carbon dioxide overabundance, and possibly deal with an excess of nitrogen oxides by treating them as nutrients. Excess volcanic sulfur compounds are automatically dealt with by large-scale oxygen sources: toxic hydrogen sulfide, carbonyl sulfide, carbon disulfide, methyl mercaptan, and sulfur dioxide are all readily oxidized by an oxygen atmosphere to sulfur trioxide, which almost instantly rains out, leaches calcium and other metals from surface dirt, and makes nontoxic insoluble sulfates. Carbon monoxide, ammonia, and hydrocarbons are also readily oxidized by oxygen and its fellow travelers such as ozone and hydrogen peroxide. Directly or indirectly, photosynthetic plants clean up a vast variety of undesirable atmospheric contaminants.

But the familiar chemical repertoire of plants that we have learned from studying Earth's atmospheric evolution and the modern atmosphere are not by any means the entire story of what biology can do for us. We can use our growing understanding of the genetic code to broaden the range of human adaptability without the necessity of waiting millions of years for natural selection and mutation to stumble upon a solution. We can imagine subtle changes that increase human tolerance of heat or cold, or permit function at unusually low oxygen levels. We can imagine adaptation of human senses to lower light levels or to a shifted spectral range: perhaps sensors to enable vision in the near-infrared part of the spectrum, based on pigments very different from those presently used in the human eye. Humans need not wait for a planet to change into an Earth. We could meet the planet halfway, adapting ourselves to local conditions.

The problem of improving human function to help us deal with alien environments need not, however, be an exclusively biological problem. The rapidly growing power of electronic and computer technology suggests a variety of ways in which these technologies can enhance human capabilities. Extremely high-throughput human-machine interfaces are already being planned. The use of electronic eyes and ears to replace those of the blind and deaf is now a near-term possibility. As these medical technologies force the development of improved interfaces between electronics and neurons, a vast range of other applications will soon follow. The integration of the human nervous system with auxiliary electronic sense organs, such as infrared eyes, is an obvious possibility. Every human may have the option of carrying an implant containing an electronic library of billions to trillions of bits of data: imagine having the entire nonfiction contents of the Library of Congress and the world's million best novels on call within your brain at all times! If knowledge is power, then I endorse giving all knowledge to the people. But no man need be an island: why not your own implanted cellular link to give you global access to the World Wide Web — with you controlling access via artificial intelligence sentries that protect you from intrusion? Why not your own internal Global Positioning System receiver to tell you exactly where you are at all times? Why not a suite of biomedical instruments that constantly monitor your health and warn you of health hazards, advise you how to maintain optimal health, diagnose illnesses, prescribe treatment, and, in an emergency, automatically call the ambulance for you?

Since about 1960, every eighteen months has seen the performance of computers double. The trend will surely continue into the future, as copper circuits, gallium arsenide chips, smaller feature sizes, and biochip technologies phase into the next few generations of computers. But as the performance of systems has grown exponentially, the cost of the "ultimate computer" has come down modestly with every new generation. When I first learned to program a computer, back in 1959, I learned a low-level language called Bell 1. The machine was a vacuum-tube computer in a room the size of a ranch house, with one thousand words of main memory. A huge air-conditioning system was needed to cool the room from the waste heat of the tens of thousands of vacuum tubes. At 8 bytes per word, and 8 bits per byte, it had 64 kilobytes of memory. Ten years later, circa 1970, I was using a CDC 3600 that was several times smaller than the vacuum-tube monster, a hundred times as fast, and with a hundred times the memory. My first-generation IBM PC, bought in November 1981, also had 64 k of RAM — on a board the size of a paperback book. It had a port for a cassette recorder and 160 kilobytes of disk storage on five-and-a-quarter-inch floppy disk drives. Today, my aging 90-MHz Pentium computer has 16 megabytes (16,000 kilobytes) of RAM, a CD-ROM drive, and a 1 gigabyte (1 million kilobytes) hard drive. *Each memory chip* in it has more than ten times the capacity of the entire mainframe I used in 1959. But now I see desktop computers on the market with 96 megabytes of RAM, DVD drives, high-resolution flat-screen graphics monitors, and clock speeds of 400 megahertz — all smaller and cheaper than that original IBM PC — and prices continue to drop. I can confidently predict that, by the time you read this, there will be better machines on the market. Computer evolution, like biological evolution, concentrates ever more power into ever smaller packages. Soon those packages will be small enough to be a part of us.

Finally, there are prospects for the wedding of the biological principle of self-replication with the exponentially increasing power of electronics. It is in principle possible to design and build tiny machines that are self-replicating. This new specialty of nanotechnology can offer devices capable of mining minerals in hazardous environments, sorting our garbage for recycling, maintaining roads and buildings, assembling larger machines, removing plaque from arteries, repairing broken or decalcified bones without surgery, and exploring alien worlds where humans cannot live—or perhaps I should say, where humans cannot yet live. Nanotechnology is potentially a very powerful tool for the taming of worlds. But the net effect of all new technology is to broaden our ability to find, extract, and use new resources: it makes everything useful. The ancient argument of utility, as taught by Lucretius, Tycho, Kepler, Huygens, and many others, asserted that everything, from nearby worlds to the most distant suns, existed for a purpose. Today we are just beginning to see how that may become true.

WHAT IS the best world for life? A world on which the world and its life have coevolved, accommodating each other. What is the best of all possible worlds? One that has been altered to our needs, and that we have met halfway by adapting ourselves. The best marriage of world and resident is consensual: both freely change for the common good. This is the direction in which we are already headed. It is best to be aware of it.

What kinds of planets, then, of all possible worlds, are well suited for eventual human occupation? A wide range indeed. There is a whole universe of opportunities out there, waiting for us; Earths and Earthlets, Earthissimos and Europas, terraformed non-Earths, planets of brown dwarfs and ocean satellites of Jovian planets around F, G, K, and M stars in spiral galaxies, globular clusters, and elliptical galaxies, and around orphan brown dwarfs in interstellar space; worlds of types known and unknown; worlds without end.

What kind of world do you want *your* grandchildren to live in?

AFTERWORD

IN THE YEAR that has passed since the hardcover edition of this book went to press, there have been several new discoveries of planets and planetary systems around other nearby stars. These systems are described briefly below, arranged in order of increasing distance from their central stars.

Three more stars have been found to have planetary companions at distances of less than 0.05 AU from their stars. The yellow G8 star HD217107 has a planetary companion with an apparent mass of 1.28 Jupiters, orbiting at a distance of 0.04 AU every 7.1 Earth days, with an orbital eccentricity of 0.14. The yellow G5 star HD187123 has a companion with 0.52 Jupiter masses orbiting at 0.042 AU, with a period of 3.097 days and an orbital eccentricity less than 0.06. The yellow-white G0 star HD75289 has a planet with a mass of 0.42 Jupiters orbiting every 3.51 days at a distance of 0.046 AU, with an orbital eccentricity of 0.054.

At slightly greater distances of 0.1 to 1.0 AU, there are three new discoveries. The orange K1 star Gliese 86 has a planet of mass 3.6 times that of Jupiter, orbiting every 15.83 Earth days at a distance of 0.11 AU, and with an orbital eccentricity of 0.05. The yellow G3 star HD195019 has a reported planet orbiting at 0.14 AU. The yellow G5 star HD168443 has a 5.04–Jupiter-mass companion orbiting it every 57.9 days at a distance of 0.277 AU, with a high orbital eccentricity of 0.54.

Finally, the yellow-white G0 star HD210277 has a companion with 1.28 Jupiter masses orbiting at a distance of 1.097 AU, with an orbital period of 437 days. Its orbital eccentricity, 0.45, is also high.

In addition to these newly discovered systems, very recent observations of the systems reported upon in this book have provided detections of two additional planets in the system of the white F8 star upsilon Andromedae. The three known planets in that system are

upsilon And B: 0.71 Jupiter masses, orbiting in 4.617 days at 0.059 AU (low eccentricity),

upsilon And C: 2.11 Jupiter masses, orbiting in 241.2 days at 0.83 AU (e near 0.1),

upsilon And D: 4.61 Jupiter masses, orbiting in 1267 days at 2.50 AU (e about 0.4).

The planet of the yellow-white G1 star 47 Ursae Majoris is now estimated to have a mass of 2.41 Jupiters and to orbit 47 UMa every 3.0 years at a distance of 2.1 AU. Finally, the planet of the white F9 star HD114762 has been found to have an eccentricity of about 0.33.

A website containing data on extrasolar planets and some brown dwarfs is maintained at the University of Paris at Meudon by Jean Schneider. The url for this site is *http://www.obspm.fr/planets*.

As with all the data discussed in this afterword and in the text, the estimated masses are lower limits, predicated upon the assumption that we are observing these systems from a vantage point in the plane of the planetary orbits. Astrometry may soon permit actual measurement of the orientation of these orbits. When such data becomes available, many of these estimates will be increased by an average factor of about 1.4. However, some rare cases, which we are observing nearly face-on, may yield masses above the limit for Jovian planets. These bodies must then be moved to the brown dwarf category.

As with the earlier results, it is very important to realize that these results are secured by an observational technique that cannot yet give us any indication of what the typical planetary system looks like. Because of the short run of observational data (a few years in the best cases), detections are very strongly biased in three ways: they favor detection of the most massive planets, those in the closest orbits to their parent stars, and those orbiting the most massive stars. The latter sounds paradoxical, since the reflex motion of the star is greatest (most easily detected) when the mass of the star is small; nonetheless, the smallest stars are so faint that observations of all but the closest are very difficult. The radial velocity technique is still so insensitive that a planet with the mass of Saturn cannot yet be detected, to say nothing of Uranian and Earthlike (terrestrial) planets.

Even more tantalizing are the reports of bodies of planetary mass orbiting about pulsars. These survivors of cataclysmic stellar explosions are detected by extremely precise timing of the arrival of long trains of radio-frequency pulses from the central star. This technique is so sensitive that planetary cores (or "cinders") of about one Earth mass may be detected. The pulsar PSR 1257 + 12 appears to have at least three planets with masses of 2.8, 3.4, and about 100 Earths (about the mass of Saturn). Pulsar PSR B1620–26 has a planet with roughly Jovian mass. Several other pulsars have been reported to exhibit pulse-timing anomalies suggestive of the presence of planetary-mass companions, but these reports have yet to be verified by independent observations. It must be emphasized that these "planetary" bodies have literally been through the hell of a supernova explosion, and can have no conceivable relevance to the question of the prevalence of Earthlike planets or of life. Some theorists suspect these bodies may even be made of material ("planet vapor") that condensed and accreted from the debris of the stellar explosion that made the pulsar.

The most recent chapter in the brown dwarf saga is the May, 1999 announcement of the discovery by two teams of astronomers of six brown dwarfs, not associated with any visible star, that are cool enough that their atmospheres contain methane. Methane, CH_4, is the simplest hydrocarbon, and a long-established component of the atmospheres of gas-giant planets. At the higher temperatures of the atmospheres of even the coolest (M-type) Main Sequence stars, methane is converted into carbon monoxide and hydrogen. These brown dwarfs are therefore transitional between super-Jovian planets and the warmer brown dwarfs discovered in previous searches. The discovery of so many brown dwarfs in the last few years is of enormous significance: these bodies are so faint that they must be very near in order to be detected. They are therefore extremely numerous, probably much more so than the visible stars.

Over the next few years, the longer span of observation will permit the discovery and verification of planets with ever longer orbital periods. Larger telescopes will be applied to this effort, possibly including dedicated planet-search telescopes in space, which will allow extension to fainter stars and smaller velocity amplitudes. The astrometric technique will soon begin not only to detect and confirm the planets already discovered, but also to provide data on the orientation of their orbits to our line of sight. The technique will allow true determination of planetary masses — not just lower limits — and permit the study of nearly face-on systems, in which the radial velocity variations are too small to detect.

GLOSSARY

abyssal plains The vast lowlands underlying the majority of Earth's deep oceans, under about 5 kilometers (three miles) of water.

accretion The gathering together of dust, rocks, and small solid bodies to make a planet.

aluminum-26 A radioactive isotope of aluminum with a half-life of about 700,000 years.

aphelion The point on a body's orbit at which it is at its greatest distance from the Sun.

apoastron The point on a body's orbit at which it is at the greatest distance from its star.

asthenosphere The portion of a planet's interior that behaves elastically, flowing in response to stresses and not breaking to produce earthquakes. See *lithosphere.*

astrometry Literally, "star measuring"; any technique of precise measurement of the position of a star, often for the purpose of discovering evidence for the existence of unseen companions orbiting it.

atmophile Literally, "breath loving" or "air loving"; a class of chemical elements that are so volatile that they are gases at ordinary temperatures. These include many compounds of hydrogen, carbon, nitrogen, sulfur, and the noble gases.

brown dwarf A body intermediate in size between gas-giant planets and main sequence stars.

chalcophile Literally, "sulfur-loving." any chemical element that readily forms chemical compounds with sulfur and thus follows sulfide melts into the core during planetary differentiation.

continental drift The motion of large lithospheric plates on the surfaces of terrestrial planets, under the combined impetus of internal convection currents and the sinking of cold slabs of lithosphere.

crust The near-surface layers of a terrestrial planet, usually characterized by high abundances of silica, aluminum, and alkali metals. By analogy, the solid surface ice layer of volatile-rich satellites and planets.

deuterium The heavy isotope of hydrogen, bearing one proton and one neutron in its nucleus and designated by physicists either as ^{2}H or D.

deuterium burning The early phase in the life of stars during which the principal source of energy is fusion reactions of deuterium, making isotopes of helium.

differentiation The process of separation of a partially melted planetary body into layers, according to the density of the respective materials, into core, mantle, and crust, including the associated partitioning of minor and trace elements into the major phases according to their chemical affinities.

dissociative recombination The process by which the neutralization of a positive ion splits a chemical bond and fires off two atoms or molecular fragments in opposite directions. An important mechanism for atmospheric loss from planets the size of Mars or smaller.

Doppler shift The shift in the wavelength of any wave, such as sound or light, caused by the relative radial motion of the source and receiver.

Earthissimo A term coined for the purposes of this book to denote a planetary body with the same overall chemical composition as Earth, but with twice the radius.

Earthlet A term coined for the purposes of this book to denote a planetary body with the same overall chemical composition as Earth, but only half as big in radius.

eccentricity A measure of the departure of an ellipse from circular shape; an eccentricity of zero is a circle, whereas an eccentricity of 1 is an infinitely elongated ellipse.

eutectic melt Literally, "good melting." A liquid, formed by heating a mixture of two mutually soluble materials, at a temperature below the melting point of either pure material. A familiar example is the mixture of salt and ice.

explosive blowoff The process of erosion of a planetary atmosphere by the explosion of massive impacting bodies.

feldspar A widespread mineral in the crusts of rocky planetary bodies, consisting principally of aluminosilicates of calcium, sodium, and potassium.

ferric An ion made by removing three electrons from an iron atom, denoted Fe^{3+} or Fe^{+++}.

ferrous An ion made by removing two electrons from an iron atom, denoted Fe^{2+} or Fe^{++}.

G3 Globs of gelatinous goo.

Galilean satellites The four large moons of Jupiter, discovered by Galileo Galilei.

gas-giant planet A planet dominated by volatile elements, especially hydrogen and helium.

giant molecular cloud A dense, cold interstellar cloud of gas and dust, massing up to millions of Suns, in which dozens of complex organic compounds are present in the gas.

giant planet A gas-giant planet (q.v.).

globular cluster A spheroidal cluster of stars, usually devoid of detectable interstellar gas and dust, and normally composed of very old stars with low "metal" content.

Grand Tour A spacecraft mission that flies by most or all of the giant planets.

H-R Diagram (Hertzsprung-Russell Diagram) A diagram relating the color (temperature) of stars to their luminosity. Named after early-twentieth-century astronomers Ejnar Hertzsprung and Henry Norris Russell, who introduced the concept of correlating the intrinsic properties of stars with each other.

helium burning The phase of life of a star, with mass greater than 0.6 Suns, in which hydrogen fusion has given way to helium fusion. Stars in this phase are called giants.

hot atom A highly energetic atom given off by a chemical reaction such as photolysis or dissociative recombination (q.v.).

hydrogen burning The main phase of the life of a star, in which hydrogen nuclei are fused to make helium. Stars in this phase are called main sequence stars.

hydrothermal vents Vents in the ocean floor through which superheated water and solutions laden with dissolved minerals are erupted from the oceanic crust into the ocean.

incompatible elements These are lithophile (q.v.) chemical elements that do not readily enter into dense iron-magnesium silicate minerals such as olivine and pyroxene, and are therefore expelled from planetary mantles.

kosmos A technical term from Greek philosophy, denoting all creation. In some philosophical schools, a kosmos corresponds roughly to the solar system; in others, to the universe as we presently understand it.

Leonid meteors A major recurrent meteor shower that strikes Earth in mid-November. The peak of the shower, only about an hour in duration, reaches enormous rates of up to 100,000 visible meteors per hour. Peak displays occur every 33 years. The next is due in 1999.

lithophile Literally, "rock loving." Any chemical element that readily forms oxides and silicates.

lithosphere The outer portion of a planet, where temperatures are so low that the rocks behave like brittle solids. See *asthenosphere*.

luminosity The amount of energy given off per second by a body such as a star. In general, luminosity is expressed as ergs or joules per second, or as watts. The luminosity of a star is often given relative to the luminosity of the Sun (4×10^{33} erg/sec), as in "a luminosity of 0.07 Suns."

magma A partially melted fluid made by heating rocks to high temperatures. Volcanic eruptions normally emit magma, not completely liquid rock.

magnitude A convenient measure of the apparent brightness of a heavenly body, derived from a traditional scale in which the brightest stars were said to be "of the first magnitude," and the faintest stars visible to the naked eye were said to be "stars of the sixth magnitude."

main sequence A continuous family of stars on the H-R Diagram (q.v.), ranging from very large, high-luminosity violet stars to very small, low-luminosity red stars, and consisting of stars that derive their luminosity from the fusion of hydrogen into helium.

MS main sequence (q.v.).

olivine A dense iron-magnesium silicate mineral, consisting of a solid solution of Mg_2SiO_4 with Fe_2SiO_4, common in planetary mantles.

periastron Literally, "near star"; the point of closest approach of an orbiting body to its central star.

perihelion Literally, "near Sun"; the point of closest approach to the Sun.

photolysis Literally, "dissolving by light"; the process by which absorption of energetic light, typically ultraviolet, breaks apart a molecule into fragments.

polyformaldehyde A long-chain molecule made of many units with the composition H_2CO, the simple organic molecule formaldehyde.

Population I Those stars found in the spiral arms of galaxies, characterized by high contents of carbon, oxygen, silicon, iron, and other heavy elements that were made by earlier generations of stars.

Population II Those stars which are most common in globular clusters, elliptical galaxies, and the galactic core, characterized by very low abundances of elements heavier than helium. These are very ancient stars, formed out of raw materials to which supernova explosions had not yet added large masses of heavy elements.

pyroxene A family of moderately dense iron-magnesium silicate minerals, consisting of solid solutions of $MgSiO_3$, $FeSiO_3$, and $CaSiO_3$, sometimes with small amounts of Al_2O_3, common in planetary mantles and in denser (basaltic) crustal rocks.

radial-velocity method An indirect method of detecting invisibly faint companions of other stars by measuring the periodic velocity changes of the parent star as it and the smaller body orbit their common center of gravity.

refractory A solid material that is stable at unusually high temperatures, either in the sense that it has a very high melting temperature or that it has a very low vapor pressure (and hence is difficult to vaporize).

runaway accretion Accretion that occurs under circumstances that cause the largest bodies to grow far more rapidly than smaller ones, and hence cause the largest to "run away" to very large size.

semimajor axis Half the length of the long axis of an elliptical orbit; equivalent to the mean distance of the orbiting body from its primary.

siderophile Literally, "metal-loving." An element that is easily reduced to the metallic state and dissolves in liquid or solid iron, and hence is extracted efficiently into metal-bearing planetary cores.

Solar Nebula The primordial cloud of gas and dust out of which the Sun and the rest of the solar system formed 4.55 billion years ago.

spectral classes of stars Distinctive groups of stars sorted according to color (temperature) and the spectral features seen at visible wavelengths. Almost all stars are categorized as O, B, A, F, G, K, or M, which form a color sequence from violet (about 20,000 K) to red (under 3000 K).

spin-orbit resonance If the rotation (spin) period and orbital period of a body are in an exact harmonic relationship, such as 1:1, 3:2, and so on, the body is said to be in a spin-orbit resonance.

stellar nebula The primordial cloud of gas and dust out of which a star and its retinue of planets and small bodies form.

subduction zone In plate tectonics, a region in which dense oceanic crust is being forced under a continental margin and begins to break up and melt in the mantle. Commonly, subduction zones define long belts of intense earthquake and volcanic activity.

sublimate To evaporate without melting, like snow in Denver.

T-Tauri stars A class of very young stars, named after the first known example of the type, characterized by moderate excess luminosity and extremely intense stellar winds.

terminator The line dividing the illuminated hemisphere of a body from its dark hemisphere, consisting of the sunset and sunrise lines on its surface. The definition becomes tedious for systems in which more than one star is present.

troilite The dense iron sulfide mineral FeS, very rare on the highly oxidized surface of Earth, but virtually ubiquitous on rocky Solar System bodies. Named after a Jesuit friar, Domenico Troili, who was an early student of meteorites, and the first to describe the mineral.

tsunami (Japanese) A "tidal" wave; any giant wave that is *not* produced by tides. Tsunamis may be produced by earthquakes, submarine landslides, or oceanic impact events.

volatile Easily vaporized. Usually applied to compounds of hydrogen, carbon, nitrogen, oxygen, sulfur, and the elemental inert gases.

white dwarf A very small, very dense star, actually ranging in color from yellow to violet, that is the product of extended evolution of moderate-mass stars.

SUGGESTED READING

THE MOST CONCENTRATED SINGLE SOURCE for historical background on the question of other worlds is Steven J. Dick's fine book, *Plurality of Worlds,* published by Cambridge University Press in 1982. Several selections from this vast and ancient literature are worth first-hand examination. One of the most elegantly written is *La Pluralité des Mondes Habités* by Camille Flammarion, originally published in Paris by Gauthier-Villars in 1864, and reissued in many subsequent editions. This delightful book was written by Flammarion in 1861, at the tender age of 19.

The idea of habitable zones around stars is explored in depth in the book, *Circumstellar Habitable Zones,* edited by L. R. Doyle, and published in 1996 by Travis House, Menlo Park, California. I have seen one recent book on the planets of other stars, by Donald Goldsmith. His volume concentrates very tightly on the recent discoveries of brown dwarfs and super-Jovian planets and on the details of how the search was conducted, and has little to say about the origin and evolution of planetary systems.

For science fiction's perspectives on the emergence of ideas about other worlds, I recommend two interesting and well-written reviews with distinctly different approaches: *Trillion Year Spree* by Brian W. Aldiss (Avon, NY, 1988), and Alexei and Cory Panshin's *The World Beyond the Hill,* published by Jeremy P. Tarcher, Inc., Los Angeles (1989).

Readers who desire a broader perspective on introductory astronomy would do well to consult any of a large number of college astronomy texts. I tend to favor *Astronomy: The Cosmic Journey,* by William K. Hartmann and Chris Impey (Wadsworth, revised frequently). For a quantitative introduction to Solar System science, with much detailed treatment of matters of origin and evolution, I suggest either my exhaustive tome, *Physics and Chemistry of the Solar System* (Academic Press, 1995 and 1997), or the shorter and more portable *Worlds Apart: A Textbook in Planetary Sciences* by Guy Consolmagno and Martha Schaefer (Prentice Hall, 1994). Many other good, small, and nonmathematical books on planets have been published, but most are sparing on the subject of origins, and nearly every one is older than the discovery of brown dwarfs.

Most of the other literature on the subject of planets of other stars is both very recent and very technical. Of that which is neither, I see little point to referring the reader to magazine articles that touch in much less detail, or from a dated perspective, on many of the same topics covered in detail and up-to-date in this book.

If you enjoyed reading about this subject, try tracking news items and sites on the World Wide Web—or, failing that, read this book again.

ACKNOWLEDGMENTS

I am grateful to Bill Hubbard and Jonathan Lunine for supplying fresh insight into the mysteries of brown dwarfs.

Many thanks to my wife, Ruth A. Lewis, for her careful reading of the manuscript and for her many suggested improvements.

I am indebted to David Egge for several of the paintings reproduced in the insert.

INDEX